A Guide to the Safe Removal of Honey Bee Colonies from Buildings

Written by:

Clive A. Stewart & Stuart A. Roberts

With a foreword by

Professor Thomas D. Seeley

A Guide to Safe Honey Bee Colony Removal from Buildings
© Clive A. Stewart & Stuart A. Roberts

ISBN 978-1-914934-50-6

Published by Northern Bee Books, 2022
Scout Bottom Farm
Mytholmroyd
Hebden Bridge
HX7 5JS (UK)

Design and artwork by DM Design and Print

Cover photograph

This is typical of the comb built by unmanaged colonies of honey bees residing in buildings. Note the emerged queen cell. Photograph by Clive A. Stewart

Dedication

Hard work, early mornings, late finishes, sprinkled with a number of successes and failures created some extremely long and tiring days, not to mention the highs and lows. These have all contributed, unknowingly at the time, to the content of this book. Without the advice and patience, of my good wife along the way I may never have got this far in doing what I love and enjoy doing today.

I dedicate this book to Polly for her continued support and motivation in all that I do, as untidy and disorganised as I maybe, and for the times I don't always tell you what I'm up to "wim wam for a duck's arse".

I would also like to give thanks to the co-author of this book Stuart Roberts for his encouragement and help in making my scribblings make sense. Not a small feat by any standard. I hope the reader has as much enjoyment in reading it as I did in putting it together.

Clive A. Stewart

Close up image of comb removed from a chimney in the centre of Solihull in August of 2020. Showing the diverse array of pollen sources in the local area. Photograph by Clive A. Stewart

Foreword

The honey bee, *Apis mellifera*, is often included in the list of 18 or so species that are domesticated e.g., cattle, chickens, and horses and that require human assistance to persist. But this is a mistake. Many colonies of honey bees live independently, having chosen cavities in trees and buildings as homesites. These wild colonies are subject to the inescapable process of natural selection; only those colonies that can take care of themselves are able to survive and reproduce. This means that the wild colonies of honey bees in a region may possess physiological and behavioural traits that give them natural resistance to various pathogens and parasites, such as American foulbrood bacteria, the chalkbrood fungus, and Varroa mites. We know that for each of these agents of disease, an important mechanism of control is good hygienic behaviour. For example, the workers in wild colonies are good at uncapping and recapping cells of brood infested with Varroa mites, and in doing so they disrupt the reproduction by these mites.

For more than 40 years, I have monitored the survival success of colonies living in buildings and trees around my hometown of Ithaca, in New York State. In doing so, I have learned that these wild colonies have survived well, both before and after Varroa mites began to infest them (in the mid-1990s). Genetic analyses of workers collected from these colonies in 1977 and 2011 have revealed that this population of wild colonies collapsed by about 70% between these two times. It is clear that this die-back was due to colonies becoming infested with Varroa and thereby infected with the deadly Deformed Wing Virus. It is also clear that when Varroa spread throughout this population of colonies, there was powerful natural selection for mechanisms of Varroa-resistance. Wild colonies are as abundant now as they were before the arrival of Varroa destructor in New York State.

Many wild colonies of honey bees take up residence in buildings, and sometimes they do need to be removed. The book that you are holding is a superbly detailed and richly illustrated guide to the removal process. I am delighted that the authors have included a thorough discussion of their experiences and recommendations regarding "Trap-out Removals" (aka"Trap-Outs" and "Cut-Outs"). Removals performed in this way can foster the survival of colonies that have persisted without help from us, and so have shown that they possess the skills needed to cope with their parasites and pathogens. These are colonies that can help us to find our way back to treatment-free beekeeping.

Professor Thomas D. Seeley – November 2022

Contents

Part 1

The natural history of the honey bee as it relates to colony relocation

Part 1. The natural history of the honey bee as it relates to colony relocation

Honey bees are thought to have come out of Africa or Asia, it is not yet known, about 100 million years ago. They coevolved with flowering plants which is why there is such a strong link between them. The species diversified over time and today we have a number of different bee families across the globe. The one that we are concerned with is the Western Honey Bee, *Apis mellifera*. They are described as cavity nesting bees and would naturally live in small hollows in trees or caves and such like.

Honey hunting was a practice where a honey bee nest would be located and then raided. This would have been typically for the honey but also for the brood combs which are very high in protein. Historically honey bees were first 'farmed' in situ in trees. This then developed to clay pipes, straw skeps, wooden boxes and today in plastic and polystyrene hives. However, it is fair to say that the honey bee has fascinated the human species for millennia.

How is it that bees find their way into your chimney or cavity wall?

This is the result of swarming.

Okay, but what is swarming?

The superorganism - we normally think of honey bees as little insects that flit from flower to flower collecting nectar and turning it into honey for our toast. However, scientists tend to think of social insects like honey bees as a superorganism. That is a group of individuals that work together towards a common goal. Any living organism has the desire to reproduce, and superorganisms are no different. The offspring of a honey bee colony is a honey bee swarm. That is to say that at a colony level, it splits into two colonies each with the right material to produce two viable colonies. At the time of swarming the colonies might not be strictly viable but should be viable units within a couple of weeks. This can be thought of as 'Colony Fission". The colony splits into two viable halves.

Bee Development

The individuals in the colony have different jobs and it is this division of labour that makes the colony so effective at what it does. As the worker honey bee ages, glandular development dictates what roles can be performed by the worker bees. For example, when their hypopharyngeal glands have developed that are able to produce royal jelly and brood food to feed the larvae. When their wax glands develop, they can produce wax and become involved in capping honey and building comb. When their sting glands develop, they are able to perform guard duty at the hive entrance and so on.

Colony Development - The Yearly Cycle

The colony clusters into a ball shape during the winter months. They bees cluster (a bit like penguins) and take it in turns to be on the outside of the cluster. In the centre of the cluster, if the queen is laying eggs, the temperature will be maintained at 35°C no matter what the temperature outside of the colony is. If the queen has stopped laying and there is a brood break, then the internal temperature will be maintained at about 20°C. On warmer sunny days (about 10°C ish) the bees may break the cluster and attempt foraging and cleansing flights. During the winter the bees maintain these temperatures by exciting their wing muscles but dislocating their wings which generates heat in the muscles. This is fuelled by the consumption of the honey that they stored back in the summer months.

The bees remain in this cluster until the ambient temperature allows them to break up. Usually, if the outside temperature is less than 10°C the bees will begin to cluster. Then as the outside temperature rises above 10°C the cluster will break up and if the weather is favourable the bees will venture out of the nest on cleansing, orientation and foraging flights.

If the honey bee colony survives the winter, then it will start to grow very rapidly in the spring because of the availability of spring nectar and pollen. There needs to be a huge amount of pollen gathered by the foraging bees in order to feed the developing brood (baby bees). It is at this time in the year, the spring, when the amount of developing brood can be greater than the number of adult bees in the colony. This can stress the colony, but most colonies get through this period successfully. The colony grows very rapidly and reaches a critical mass (as a superorganism) where the colony believes that it can split into two viable colonies.

The colony continues to grow in size both in terms of the number of adult bees and the area of comb that the colony has available for both brood (egg laying) and food storage (honey).

It is usually at this stage, during the colony expansion, that swarm preparation begins. I'll come back to swarm preparation shortly.

What is Swarming?

Continuation of the species - Swarming, as I have already written, is the colony splitting into two separate colonies. But what is a swarm? A swarm is usually about 30,000 adult bees taking to the air on their way to find a new home. The swarm consists of 1 queen (usually) which is the mated, fertile, egg laying queen from the nest. She is accompanied by mainly young worker bees carrying honey in their honey stomachs. The swarm issues from the colony when the first of these replacement queens reaches the pupal stage of development (when the first queen cell is capped). The bees leave the existing colony and will typically cluster in a tree within about 20 metres of the original colony. It is here that the final voting will be made as the bees decide where to make their new nest. This decision making has been thoroughly researched and documented by Dr Tom Seeley (Professor, Cornell University) in his book 'Honeybee Democracy'.

The bees will typically stay in this swarm cluster for a few hours up to a couple of days. When the decision-making process has concluded the swarm cluster breaks up and the bees fly off to their new nest site led by the scout bees that found the site.

Very occasionally, the bees either fail to come to a decision or there are no suitable cavities within flying distance and the colony remains in the swarm cluster too long. This results in the young bees building wax comb in the tree. The fertile laying queen will then lay eggs in this comb. Once there is an established brood nest in the tree then the bees will not abandon it and the swarm cluster, now more accurately called a colony is doomed to perish. This is because they have no shelter or protection from the elements or predators that a cavity would normally provide.

Swarm cluster in a tree by Stuart A. Roberts

Swarm cluster on a fence by Clive Stewart

Swarm cluster in a tree by Clive Stewart

Swarm cluster in a tree by Clive Stewart

The Cause of Swarming – The Theories

The process of swarming is complex and not fully understood. There are several theories behind the causes of swarming and none of them are fully able to explain the cause. However, it is likely to be a combination of the theories. The main and most highly accepted one is the queen pheromone theory. Queen pheromone was discovered by Colin Butler at the Rothamsted Research facility in the 1950s. When a queen is newly emerged, or within a few days, she is producing 'Queen Substance' the queen pheromone at its maximum rate. This rate of production diminishes over time so that after about 1 year the queen is producing the pheromone at about half the rate she was when she was newly emerged. It is thought that a lack of 'Queen Substance' is one of the triggers for swarming. Therefore, a queen in her second year is more likely to swarm than a queen in her first year, according to this theory.

The pheromones from the queen are passed from bee to bee in the colony by antennation and trophallaxis, that is touching each other with their antennae and sharing food between bees, both of which are commonplace. Through these mechanisms the pheromones pass through the whole colony very quickly, usually within minutes. If the queen is not producing as much pheromone and the colony is expanding rapidly, it is logical that there might be a dearth of pheromones at the edges of the nest. It is this that is thought to initiate the swarming instinct.

Once the bees have decided to swarm, a sequence of events begins that culminates in the production of a number of new queens to replace the existing, fertile, egg laying queen in the colony. I'll expand on this later.

Other older theories, based more on observation, consider an excess of nurse bees (young bees that are feeding the brood) as a potential trigger to swarming. Also, as the colony builds up in the spring, the colony can become more congested, and it is thought that congestion can be a trigger for swarm preparations.

Swarm cluster in a tree by Stuart A. Roberts

The Swarming Process

I have already stated that swarming is colony reproduction. I mentioned earlier that the honey bee colony will rapidly expand in the spring. If the conditions are looking favourable, then the colony will 'decide' to reproduce. Favourable conditions would be good weather, a strong nectar flow and a low disease load.

The colony will begin by producing drones and this is usually that first sign that the colony is investing in reproduction. The colony will need to replace the existing fertile laying queen. I wrote earlier that the existing queen will leave with the swarm. However, the colony will not take a chance on just a single replacement queen but will make a number of queens to ensure that the chances of getting a replacement fertile queen are high. The number of queens that the bees attempt to make varies. A colony of 'swarmy' bees might make tens or even a hundred queens. More normally, a colony might make about 20 new queens.

How do the Bees Make a New Queen?

A queen is made from a normal female egg laid by the incumbent queen. The resulting larva is then fed a protein-rich diet of royal jelly for the whole of its life as a larva. Recent research has shown that it may be the quantity of food fed rather than a different diet. It postulated that queen larvae were fed much more than worker larvae and that was enough to produce a queen rather than the higher protein diet.

The photo shows normal worker cells and an emerged queen cell
(the queen has emerged from the cell) – Photo by Stuart A. Roberts

Worker bees are female too, but they have a less protein-rich diet after day three (as a larva) when their food is switched from royal jelly to worker jelly.

The other difference is the orientation of the cell. A worker cell is in the plane of the comb whereas a queen cell hangs down vertically from the comb.

Photograph, by Clive Stewart, showing emerged queen cells in comb recovered from a building.

The new queens develop to maturity in these queen cells. Sixteen days after the egg was laid the adult queen will bite her way out of her queen cell with the help of the workers.

On emergence the 'superorganism' has a decision to make. This newly emerged virgin queen could leave with an afterswarm, sometimes called a cast. Or the alternative is that the newly emerged virgin queen hunts down her rival sister queens (possibly still in their cells) and the queens fight to the death. There can be only one reigning queen under normal circumstances.

However, 8 days earlier, the main swarm or 'prime swarm' will have left the colony. This is triggered by the new queens entering the pupal phase of development. This is when the queen cells are capped. At this time the swarm leaves with the incumbent fertile laying queen and about three quarters of the adult bees, the majority of which are young bees. This is because when the swarm gets to the new nest location, they will need to build wax and it is the younger bees that are the wax builders.

Leaving the nest

At the point when the swarm issues from the nest the colony has already been sending scouts out looking for new nesting sites or new cavities. However, the bees in the swarm cluster will still need to make a final decision on the ultimate nest location.

The bees rush out of the original nest and gather close by, within 20 metres normally. The bees cluster into an elongated ball shape (think American football on end) and will remain in this cluster around the queen until the final nest location has been agreed.

Tom Seeley's work is fascinating and tells of the bees scouting out new nest sites and then voting for them via the medium of the 'waggle-dance'. The scout bees communicate the direction and distance of the newly discovered cavities by dancing on the surface of the cluster. The scout bees will find a cavity somewhere and investigate it. If they decide it is worth telling the other bees about then they come back to the swarm cluster and dance on the surface of the cluster trying to recruit other scouts to look at the cavity that they have found. When there is a consensus, when all of the scouts are dancing for the same cavity then the cluster will again take to the air and fly to this agreed nesting location.

What this means practically

This means that honey bee scouts will find and investigate cavities. Not all of them will be suitable and some will be more desirable than others. For cavities that are deemed suitable by the bees this could ultimately mean occupation by honey bees. This is how a swarm of bees will locate and end up in your chimney or cavity wall.

Typical locations for swarms to investigate or go to

Any cavity that meets the bees' needs is a possible nesting site. Tom Seeley investigated the bees' preferences for size of cavity, entrance size and direction of entrance amongst other things. He concluded that the bees prefer a cavity that is between 40 and 70 litres and is significantly off the ground by a few metres with a small entrance that faces in a south easterly direction.

This would be their preference. However, when the ideal isn't available then any cavity may do. I have seen bees in chimneys (usually blocked up disused ones), the cavity walls of buildings, the ceilings of buildings, the roofs of buildings. I have personal experience of bees in an old post-box that was part of a monument at the National Memorial Arboretum at Alrewas in Staffordshire.

Swarming – Is it Bad?

Swarming is seen as a bad thing from a beekeeper's point of view. Firstly, and perhaps selfishly, losing half of the bees to a swarm means that the beekeeper will not get a honey crop. To get a honey crop you need a large foraging force and losing half of the bees is not conducive with a large foraging force.

Secondly, as beekeepers, we try not to upset our neighbours by causing them issues with our bees. Having a swarm cluster hanging in one of their trees in their garden is one thing, perhaps a slight annoyance as you gather it back again. However, your bees taking up residence in their chimney will have a much greater effect on their neighbourly demeanour. Especially when they find out how much it is likely to cost them to get the bees removed.

It is the beekeeper's responsibility to 'keep' the bees together as a colony or to manipulate the bees so that the swarming urge is removed. They are many ways to do this, and they will not be covered in this book. This book is concerned with what to do when these measures fail.

However, swarming is a natural behaviour for the honey bee. It is a good sign that the colony is growing and is prosperous and usually healthy. It is a indication that the bees are behaving normally, which can only be a good thing.

The behaviour of bees that helps removal or swarm collection

Communication in honey bees is mainly by pheromones and vibration. The queen, as I mentioned earlier, produces a cocktail of pheromones called the queen substance. This is key in holding the colony together and the worker bees know very quickly if the queen has been removed.

In a swarm situation, if the swarm leaves the nest and forms the swarm cluster in a nearby tree, it is the queen that keeps that cluster together through the pheromones she exudes. If a swarm leaves a nest and, for some reason, the queen does not make it to the swarm cluster then the swarm cluster will very quickly break up and the worker bees will return to the original nest having abandoned the swarm attempt.

If the queen can be identified in the swarm cluster and placed in a box (think normal cardboard box – large Amazon delivery) then the worker bees will search for her and will want to be with her. This forms that basis of most swarm collection strategy. If you can find the queen and put her in a box, the worker bees will follow. You can then remove that box with the bees in. Obviously, it's not quite that simple but that is the basis of the technique. See the section on swarm collection later in the book.

Part 2

Health, Safety, and Legislation

Part 2. Health, Safety, and Legislation

The one thing that must be remembered when carrying out any bee removal work, is that one sting can be fatal to the wrong person, and that person is unlikely to know that they are anaphylactic. Anaphylaxis is a severe allergic life-threatening reaction. It is estimated that 5% of the population are believed to have the potential for an anaphylactic reaction to bee stings, and it is, therefore, imperative that health and safety is given some serious consideration.

Some removals will be much simpler than others, but no one removal is the same, even in such situations as when you are dealing with two chimneys on the same building.

There are a number of acts, and individual parts of legislation associated with the removal of all colonies within buildings, and all factors of each individual situation should be taken into consideration before any work is to commence. We are not going to go into every little detail of each piece of legislation within this book, the Health and Safety at Work Act 1974 has 84 sections alone. We will identify those that are most important and recommend that the reader seeks further advice and suitable training in making sure that they are working within the law, and safely.

Typical wild comb found in a ceiling void – photograph by Clive Stewart

Health and Safety at Work Act 1974

A detailed survey should identify any hazards for the provision of any Risk Assessment & Method Statement (RAMS). The RAMS is not just a piece of paper that identifies the risks and how you manage them within your particular work area, but it is also a program of work and should identify what you will do in the case of any emergency, and what order actions are carried out in. Whilst all parts of this act are applicable to all that are carrying out works on any property be it domestic or commercial, and regardless of whether or not it is free or charged for, there is one section in particular that is exceptionally important to the person undertaking any removal. **Section 8** refers to the safety of those close by, this does not necessarily mean just those operatives working for you, or with you, but also includes the safety of the general public. As you can imagine, controlling eighty thousand bees in the height of the season during a removal can be extremely difficult, and to secure the safety of others during the operation can be even more problematic.

People are inquisitive, and everyone wants a picture of the bees and the bee's hero. It is also inevitable that many questions are asked. Which is great for the profile of the bee remover, but not for that one person that chooses to ignore all the warning signs and gets stung, even after the seventh verbal recommendation to step back to the safe zone.

Where you can, you should always notify those residing closest to the operation of what is going on, when it will be safe, and where the safe zones are. It is good practice to allow people to come and ask questions prior to the start of any work and leave your contact details with those that have an interest in the operation. Those with children are usually more anxious about proceedings, and it's a good recommendation to advise them that they go and visit grandparents, or other close relatives for the day. I also offer to take pictures for the client and advise the curious to engage with them for copies. Just highlighting this one individual section shows that there is an enormous amount of preparation needed even before contemplating how the removal work is to be done.

As part of *due diligence* in the understanding of the Health and Safety at work Act, you should give consideration to the working at height regulations of 2005, and confined spaces Regulations 1997. Both of these have bearing on the work to be carried out in a lot of situations involving the removal of colonies, as most honey bee colonies are found at height or hard to reach places. It cannot be stressed enough that a detailed survey is so important to the success of a safe removal. Not all situations at height will require the need for access equipment and can sometimes be executed from inside with feet firmly on the floor of the level at which the removal is to take place. This often has the added benefit of reducing the financial burden for the client, and usually takes away the need for specialist access equipment.

This is where the Confined Spaces Regulations 1997 may come into play, particularly in scenarios where colonies are inside small loft spaces. Thankfully these are few and far between. For these types of situations, it is imperative to work in a team, and have an emergency evacuation plan. Not only for yourself, but for the emergency services getting you out should something go wrong. If this has not been thought through it could delay your rescue and put your life at risk.

Ladders should never be used as a working platform, and not only is the individual above risking life and limb by doing so. The balancing of the ladder on a single unsecured scaffold plank is a recipe for disaster. The scaffolding clearly also does not meet with any sort of safety standards.
Photo by Clive A Stewart

When situations do require working at height, as is often the case, careful consideration should be made as to how this will be done. Ladders are not a working platform and should only be used as a tool to access a working platform, or short period work. They should never be used to carry out the entire removal. Dealing with thousands of angry stinging insects on a ladder, whilst trying to dismantle a building is simply asking for trouble.

There are a number of much safer ways this can be achieved, the simplest, and for me the safest is by scaffolding. The biggest positive for this method is that it helps to put barriers between you, the ground, and your safe zones. Whilst you are attracting the attention of the disgruntled attendants of the colony that you are removing, should you feel that the situation is a little overwhelming, and believe me on large colonies, at the height of summer, in smouldering heat, and on top of a roof with no shade it can be a little uncomfortable.

You may find the need to come down out of the way, and coming through the individual levels of the scaffolding, be it a custom built one or a module scaffold tower, the importance of those breaks soon becomes clear, as the bees reduce in number as you go through each floor.

Note: The use of kickboards, safety gates, and the pulley wheel. Photo by Clive A Stewart

Scaffolding should be erected in a way that allows as much space and access as feasibly possible. Scaffolders aren't beekeepers, and it's often a requirement that bee suits are supplied as part of their PPE. This goes for anyone that will be working with you prior to the removal who is going to be in close proximity to the bees.

Always be sure to use a qualified scaffolder for this type of access, and if you are using a mobile scaffold tower you should hold the correct qualification for this too. The down side to scaffolding is that it can be costly. However, so are the results of falling from an unsafe ladder.

An alternative is the Mobile Elevated Work Platforms (MEWPs) also affectionately referred to as the cherry picker. This is not my preferred method of access, as there is no ability to create a barrier when using them, and safe zones become more difficult to implement. This is because bees will follow you down to the ground, as they have an unbroken line of sight of you in your little basket.

Another negative for the MEWPs is that, as a safety feature they are only allowed to carry a certain weight, this varies from model to model, and when taking a building apart you add waste materials to your basket, the safe working load of the machine is then compromised, and it will stop working until the safe working load is re-established. There is also very little room in the basket, and this can also be troublesome when there are bee-vacs and nuc boxes to manage, add power cables & hand tools, and you can soon be struggling to move freely in your little basket.

As you can imagine this can be somewhat annoying during a removal, with the need to keep going down to the ground for various reasons, this inevitably prolongs the work. However, MEWPs do have their uses in certain situations, but should be given careful consideration. It should also be remembered that you will need to have the appropriate qualifications in order to legally operate a MEWPs in the UK.

A well planned and thought-out use of the basket of a track mounted MEWPS.
(Photo by Graham Lauderdale Pest Pro UK).

It is a legal requirement to have the appropriate qualification to operate any MEWPS in the UK.
(Photo by Graham Lauderdale Pest Pro UK)

Key things to remember when assessing any removal is what you will do if it goes wrong. Ensure you identify all potential hazards on your initial survey. Once a cost has been agreed I will sometimes, and particularly with commercial clients, go back to site to do a site safety check. I will often do this with the sites' safety officer, who may not have been available on my first visit.

Every part of that removal has the potential of going pear shaped (wrong) until the bees are on the hive stand and sited in their new location. So, it is important to work safely at all times for both those around you, and yourself.

Buildings Act 1984

Careful consideration should be made when executing a cut out, as you are likely to be taking apart someone's commercial property or a much-loved home. Those homes and properties are given a level of protection under The Buildings Act 1984. This is the act under which the buildings regulations 2010 work, these are regularly updated, and amended, the last amendment being in 2015.

These regulations are effectively the standards which any design, construction, alterations, or repairs, of new or existing buildings should meet. Whilst some works, such as removing a panel of plasterboard, will not generally fall under this due to the nature of specifics of individual situations, others such as the dismantling or alteration of a chimney will, and these have a specific document for their criteria.

If ever in any doubt of what can or cannot be done regarding any given bee removal situation, you should consult a professional builder or architect, who should be able to advise the need to involve the local authority planning office or not.

In the case of listed buildings, their preservation is paramount, and in most cases, but not all, supersedes building regulations. Where these buildings are concerned, local authority conservation officers should be consulted. This is the responsibility of the property owner, and proof of listed building consent should always be sought before commencement of any works.

The cost of not checking that the property you are working on is a listed building can be an expensive one to correct, and it's very simple in today's connected world to search the listed building databases either by local authorities, or the Historic England website to confirm any concerns regarding the buildings listed status. Failure to get this correct can result in unlimited fines, and or imprisonment.

Colony behind Plaster Board – photograph by Clive Stewart

Bees in chimney at the listed building Great Massingham Norfolk. Photo by The Apiarist Atelier

View of the bee colony in the chimney at Great Massingham from beaneath
Note the Lime lining of the chimney walls. Photo by The Apiarist Atelier

CDM Regs 2015

Control and Design Management (CDM) Regulations 2015 are the belt and braces to both the health and safety at work act and the building regs. This has basically been achieved by the HSE outlining the roles and responsibilities of those individuals involved in any one project. If you remove or take apart any section of a building to gain access to a colony you will fall under these regulations, and it is strongly advisable that you get some training in this area and fully understand them.

Control of Asbestos Regulations 2012

Whilst this particular set of regulations is in part applicable in due diligence to both the Health and Safety Act, and the Building Regulations, I feel it is valuable to emphasise the importance of this particular subject on its own. The words set out in these paragraphs are not a substitute for any approved training, and only serve as a reminder to the reader of the importance of the need for suitable training in this field.

Asbestos related diseases amount to approximately 4500 thousand deaths per year. The inhalation of the microscopic fibres has devastating effects on the respiratory system causing lung cancer, asbestosis, mesothelioma, and pleural plaques to name a few. As these fibres can also be digested, they are not just affiliated to the respiratory system alone and can cause similar issues to other organs throughout the body.

Asbestos cowling potentially fitted to asbestos flue inside the stack. This will require sampling & testing by those qualified to do so, to determine which type of asbestos it is, and identify what course of action is required in order to safely remove the bees. Photo by Clive A Stewart

Asbestos is a group of fibrous mineral silicates mined from the earth and was used in the construction industry up until 1999 when it became banned in the UK. However, stocks were not depleted till around 2005, and buildings up to this age may still contain elements of asbestos materials. Asbestos can be found in many parts of the building including ceiling tiles, decorative textured coatings such as Artex, Insulation boards, and soffit fittings. As you can imagine there is potentially a high risk of the bee remover coming into contact with asbestos on a regular basis.

The minimum requirement for any prospective bee remover should be an approved asbestos awareness course. This course will not only help to keep you safe but will highlight where you are likely to come across asbestos, and what to do if you should come across any as part of your survey or work.

As part of the Control of Asbestos Regulations 2012, the management of commercial buildings must have an asbestos survey and register for any works carried out in those buildings. This will include domestic buildings of local authorities, and housing associations as these form part of their commercial operations. Asking for this documentation should form part of a detailed survey for any honey bee removal that is to be potentially carried out on such buildings.

In a domestic situation these are not usually in place, unless as already stated it is part of a local authority or housing association. If you suspect that asbestos is present during your survey in a domestic property, you should stop, and advise that the offending area or items are tested for the presence of asbestos. Tests will identify types of asbestos and determine what course of action is to be taken. As a minimum the bee remover should be asbestos awareness trained. These particular removals can become complicated. However, it is not impossible for bees to be removed from these sorts of situations. Though, the client should be made aware of necessary extra work needed to prevent fibre release, and the removal of hazardous waste from site.

Food Safety and Hygiene Regulations 2013

This set of regulations is quite possibly the most breached regulation of all where bee removal is concerned. I despair and see the transgressions within the beekeeping and even the pest control communities, via many social media groups, and forums. Where the triumphant bee remover is merrily jarring his bounty from their recent bee rescue the day before. I understand that it can be all too tempting when you open up a cavity, in a wall or ceiling, and see it crammed with fin upon fin of honey filled comb. The urge to preserve every last drop for prosperity or profit is too overwhelming for some.

Sadly, this has the grave potential to do more harm than good. Not only will this honey be contaminated with the multitude of fine dust particles from the cement, plasterboard, or treated wood fibres, which were created during the process of opening up the colony to begin the removal, but the vast majority will not have been capped. This is likely to have higher than acceptable levels of water content, which in turn will cause the honey to ferment in time. It is also worth noting that some colonies may have been treated with insecticidal pesticides in the past, which also has the potential to contaminate the honey removed from any long-established colony.

Clear indication of contaminated Honey Comb, and why you would not use any honey for food from any cut out.

There are filters to remove the large particles, but they are not fine enough to remove everything. In short it is much safer to dispose of through melting processes to retrieve the small quantities of wax than it is to allow it to be put into the food chain. The comb that has the potential to have been contaminated directly with pesticide should go for incineration, and you will need to discuss this with your waste management company.

Some bee removers who manage the bees post removal simply feed it back to the bees, as a way of disposal. Again, there needs to be assurance that there are no historical pesticides involved. It should be noted that if this is something you wish to do, you should only ever feed the honey to those bees from which it came. To feed other bees with the honey from a different colony can heighten the risk of those bees contracting the notifiable diseases EFB, or AFB, and should be avoided at all costs.

Bees Act 1980 and Pest Control Order 2006

Notifiable bee diseases are covered by this act, and in particular EFB (European Foul Brood) & AFB (American Foul Brood). Technically speaking if either of these diseases are suspected to be present during a removal, then all work should cease until a bee inspector is able to confirm it is indeed the case.

However, firstly in over a decade of undertaking bee removals I have not yet come across such a situation, secondly this may prove difficult when you are working against the clock, or have a roof open to the elements.

Putting this scenario aside, it is advisable that all established colonies, regardless of where they have been taken from, should be placed into a quarantine apiary. This apiary should be identified as a quarantine apiary with your local bee inspector, and if you use the online recording system through the NBU's (National Bee Unit) website "BeeBase", it is recommended that you identify it there too.

Healthy colony behind plasterboard –
photograph by Clive Stewart

Both of these diseases are highly contagious, and easily spread to healthy colonies. By placing the removed colonies into isolation, it prevents the spread of the disease to any healthy production colonies. An advantage of identifying a specific out apiary as a quarantine apiary, is that if there is an outbreak in the area, the local bee inspector has a starting point from where they can begin their search for any potential trigger point, and potentially trace it back to the point of origin.

If you happen to be a pest control operative reading this book, and do not wish to involve yourself with the beekeeping side of this work. It is recommended that when passing on bees to a beekeeper, you should ensure that they have registered themselves with the NBU.

It is good practice to always make sure that the colonies you pass on are in the best condition possible too. Bee welfare is covered in Part 4 later in the book.

I regularly hear of colonies being placed in the care of beekeepers from pest control companies that resemble being put through a washing machine. Bees in this condition rarely survive, and not only will they have to struggle cleaning themselves, but they will have endured enormous amounts of stress during the removal. This prolonged stress is a potential trigger for EFB, and as such enforces the justification for the need to quarantine any removed colonies.

Do's & Don'ts

In summary of the above legislation, and health & safety requisites, but not limited to, here are some do's and don'ts of bee removal work.

Don't: Get complacent, it is not a clever thing to attempt to show the public how calm and friendly bees are during a removal. Because from time to time they can be displeased with your actions, and you really want to avoid that trip to the hospital that gets you noticed as the elephant man on social media, or the newspapers.

Do: Always wear your bee suit!! Gloves? gloves are good too.

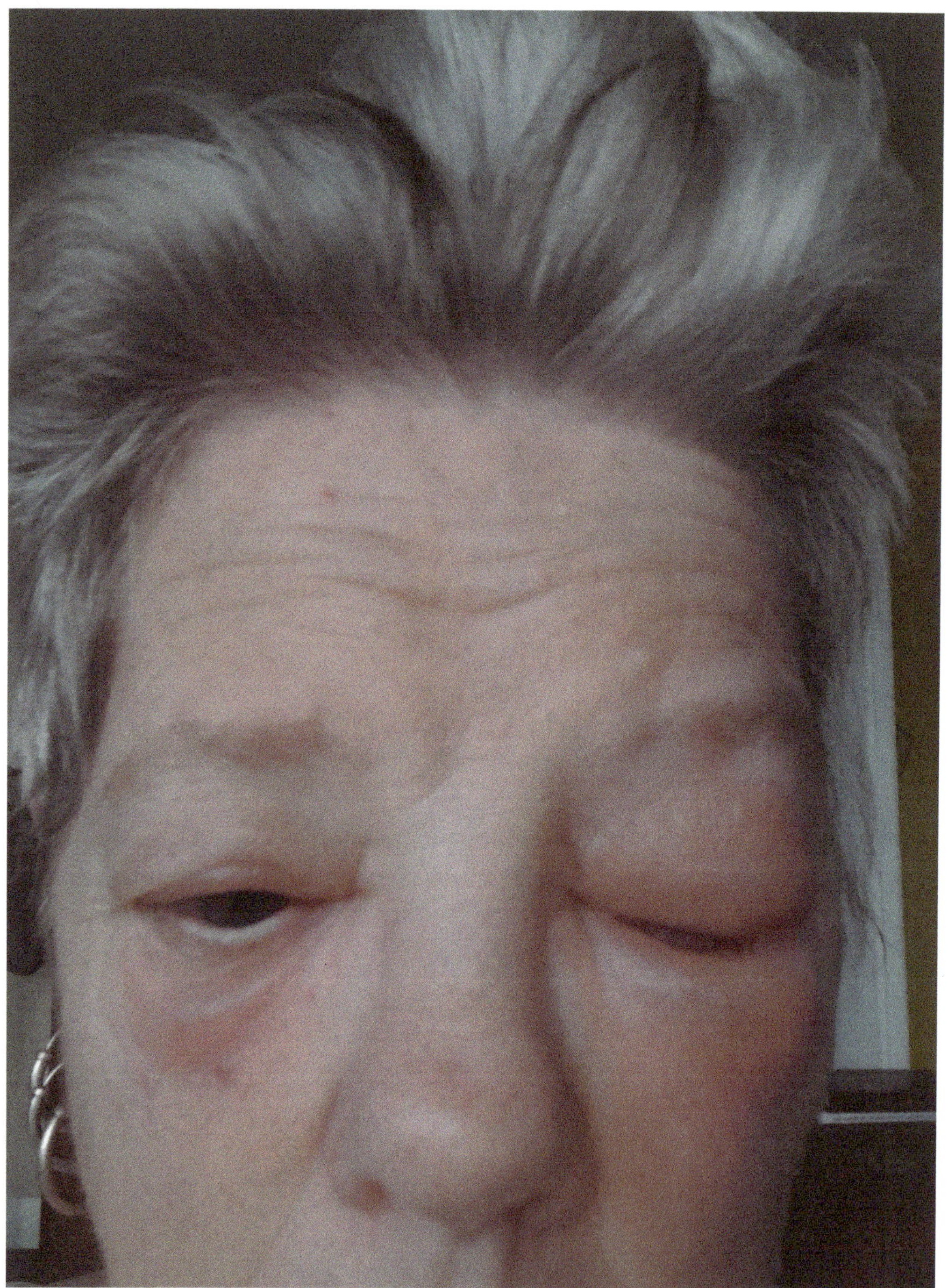

Stings to the face can be quite serious, and the wearing of the correct PPE should never be under estimated. Photo by Pauline Stewart

Don't: Complete a bee removal from off a ladder. If you are unfortunate enough to get bees on the wrong side of your veil (inside), and it happens. You will likely be doing your best impression of an Orangutan at a rave, and the top of a ladder is not the place to be starting this. Remember a ladder is not a working platform! It is a tool on which to reach a safe work platform.

Do: Use scaffold, or a MEWP's, (if your risk assessment allows), as the safe working platform.

Don't: Use or sell the dirty contaminated honey from the removal at the local farmer's market, as this is likely to lead to complaints about the grit in it. Plasterboard is not a presentable look in your honey either.

Do: Dispose of it safely, and responsibly. Whichever way you do, be sure that no unrelated bees can feed on it.

Don't: Become a statistic for asbestos related deaths. There really is no need to take risks where asbestos is concerned.

Do: Complete an asbestos awareness training course, as a minimum, and find a reputable asbestos removal firm to work alongside.

Don't: Go in blind, to attempt a removal without a full, and thorough survey, is to ask for trouble. It is very easy to perceive from a photo in an email the simplicity in a potential job, and in an attempt to save yourself some work, and your client some money, executing the work without viewing it first is simply just a foolish notion that should not even be entertained. You cannot see both sides of a wall from a photo.

Do: Always carry out a survey. Yours, and your client's safety depend on it. Surveying the property will also save time and money in executing a successful removal.

Insurance

Home Insurance Cover

I feel that it is worth mentioning a little bit about insurance from the client's perspective, as this sort of work can often become costly, particularly when height is involved due to costs of access equipment, and the question of whether or not the clients home insurance will cover the costs is a common one.

At the time of writing, there are currently no home insurance policies in the UK that cover bee removal work, regardless of any public health risks that may be involved. This is in part due to the fact that the insurance companies since around 2012 decided that bees would no longer be classed as pests, and only recognised them as beneficial insects.

However, some, but not all home insurance policies, may cover the reinstatement works. It must be pointed out that this is dependent entirely on the policy provider, and it is fair to say that the responsibility for checking whether or not a policy is valid for such claims lies firmly with the client. Should your client expect you to work under their policy claim, be sure to have secured proof to this effect prior to the commencement of any works.

PS. Since starting this book in 2022 one home insurance provider now has a policy to cover bee removal. However, it is only for a set amount, and is only available on their premium policies. I would advise that you recommend to your clients that they check with their home insurance provider.

Insurance for Business

So, you want to be a bee remover? Well one of the first things that you should consider doing is to secure some PLI, (Public Liability Insurance). Public liability in short covers compensation claims should you be unfortunate enough to have a member of the public seek a claim for injury or damage. At the time of going to press there are no direct policies to my knowledge that specifically, and solely covers the bee remover. This is simply because it is not recognised as an industry at present. It is also in part down to the many requirements of specialised skills from other industries that make it too difficult to insure under one sole industry.

However, it is commonly recognised as activities of those of a pest controller, and this is likely to be the closest industry policy available. You must inform your insurance broker, or policy provider that you are carrying out bee removal work as part of those activities. Insurance policies refer to this as "special mention" and will likely need to be seen on the insurance documentation when working for commercial organisations. You must make sure that you will be covered for every aspect that you believe you will work under with regards to your activities, failure to do so could find you liable for any incident costs yourself.

If you are also going to cover the building, reinstatement work you should also potentially seek additional PLI for those activities too. In the same respect as the pest control cover, those seeking PLI in the building side too, should be certain that their policy covers all those aspects of their activities that they are likely to come up against. Any specialist requirements for particular industry skills such as listed building works, and stone masonry should be identified in "special mentions".

The same should also be applied if you are to work with asbestos, as this is generally not permitted under normal cover of the other two industries already mentioned, and you may find that you will need to find independent cover for this too.

The question, "What do you advise?" is a popular one from clients. As part of any PLI, Professional Indemnity (PI) is recommended to form part of that policy too. This covers compensation, if your client claims for a mistake in your work, or a loss resulting in poor advice. Almost all commercial clients will insist that this is in place, and in my opinion, it is a foolish thing not to have, and particularly with regard to the activities of a bee remover.

If you have any employees that are not closely related to you, it is compulsory, and in fact a legal requirement to have in place EL (Employers Liability). This will cover compensation claims for illness, or injury caused by those activities which are associated with working for you. If found to be uninsured for this specific requirement, you can be fined per day, for every day that you were uninsured for.

Part 3

Assessment of Colonies in Buildings

Part 3 – Assessment of Colonies in Buildings

Surveying a colony for removal is one of the bee removers' most important jobs. We have already highlighted the need to identify health and safety implications during the survey. Noting the age and condition of any given building goes part way into this. How we accomplish this is outlined in this section, and we will look at, and consider some of the equipment available, in order to paint a picture of what is going on.

The reader must take into consideration that techniques, and methods are changing regularly, and I suspect even by the time we have finished this book there will already be a better or simpler way of doing something that I have written about in this section. A quick history behind the development of some of the equipment used shows this eloquently. All industries evolve over time, and bee removal is no exception. So, it is a good idea to regularly keep up to date through appropriate training and education.

Visual inspections

All surveys generally begin outside with an overall look at the situation and the property. This helps identify a number of things. Firstly, it gives an indication as to how healthy a colony is. Strong, confident, and lively activity is a fair indication that the colony is in relatively good health. Slow, lethargic, bewildered activity, and the possible sight of bees crawling around on the floor appearing lost, would indicate a colony that is not likely to be in such good health.

We can also gauge the colony's behaviour from this position and assess how defensive they are of their home. Inspecting the outside should not just involve the bee issues. A good bee remover will also be assessing the condition of the property, styles of architecture, and types of materials. These observations will give a good understanding as to the age of a building, and any serious defects they may need to be aware of, or address prior to commencement of any removal work. This in turn will have an effect on the client interview, which is as important as the survey itself.

Knowing how long the colony has been in situ is always a good start to the conversation. However, this cannot be relied upon entirely, as the client may have only moved in at Christmas, and therefore is likely not to have been made aware of its presence until the first warm day in Spring. Or they may not have noticed the swarm arriving in the previous year's summer.

Having completed the external examination of the building, you should then consider focusing your attention on the internal inspection. Here you will not necessarily be watching bee behaviour or assessing the health of the colony. Although if the colony is high up you may want to use a viewing point from inside an adjacent window nearer to the colony entrance if it is feasible. There may also be some internal bee activity that is likely to be causing stress, and alarm to the occupants of the building.

Opening a window and blocking holes will help subdue this sort of situation. The objective of carrying out an internal inspection will be to get a better idea of exactly where in the structure the colony is located, and gauge how large or small the established colony is likely to be. It will also give a better indication as to how best to remove the colony from its position, and whether or not an external or internal removal would be best.

Obviously, dependent on where the entrance is externally, will depend on where you will begin your internal survey. You would not be looking at the kitchen ceiling if the bees were entering and exiting the chimney pot. Here it is important to have the right equipment to carry out the necessary investigations. Listening devices, thermal imaging equipment, serpentine inspection cameras, and other specialist equipment helps paint a picture of what is going on and how best to remove the colony.

We will look more closely at some of the important items in the bee removers survey bag further into this chapter. Documenting your findings is equally important as the survey itself. This not only helps you complete an accurate risk assessment and method statement of the works that you are going to be undertaking, but it also shows the client the extent of the situation within their building and helps to identify the most economical way to remove the colony.

A) Identification, B) carry all, C) chimney camera and monitor, D) thermal imaging camera, E) endoscope camera, F) stud/services detector, G) electronic measuring device, H) report pad, I) stethoscope - photograph by Clive A Stewart

The use of thermal cameras

The development of thermal imaging has a history that goes back as far as 1800 when Sir William Herschel discovered infrared radiation, on which the modern-day thermal imaging cameras work. In 1929 the Hungarian physicist Kalman Tihanyi developed the first thermal camera for antiaircraft defence following World War one. It developed further in 1947 when the US military along with Texas instruments created the first line scanner, where a single image was able to be produced. However, the images took one hour at a time to produce. Further developments between 1970 & 1978 Brought thermal imaging into the firefighting industry. It wouldn't be till the early 2000's that it would become affordable, and readily available to the general public.

A thermal imaging camera today detects varying levels of infrared light. The hotter an object is, the more infrared radiation is produced, the camera is able to detect this, and reconstruct it into an image that we can see. It is no understatement that thermal imaging has come on in leaps and bounds in recent years. So much so it is now as compact as a camera on your phone, or an adaptation that you fit to it.

Thermal imaging cameras are incredibly light, and amazingly accurate. Modern day smartphones with this equipment built in are extremely useful as other Apps allow the downloading of images to documents, and emails straight to the internet without the need for cables for reporting purposes. Many of the purpose-built devices also allow this option, but usually require cables to be connected to devices or memory cards to enable them to do what the phone devices do already.

A typical survey photo – visible light vs infra-red light – Photos by Clive Stewart

The thermal image of the photo above – Photograph by Clive Stewart

What must be remembered, is that they are not to be solely relied upon to complete a detailed survey; they simply detect the heat created by the main body of the colony, it does not show all of the combed area, and can convince the inexperienced that a colony is smaller than it actually is.

This is never more evident than in the Spring as colonies will be tightly clustered keeping warm in the centre. It is therefore important that you ask the client for as much information as possible to help gauge the colony's age and size.

There are a number of things to be aware of when utilising this piece of equipment for the purpose of locating the colony, and that is that the colony is not always exactly where the entrance is. Sometimes it can be located as much as 2.5m away from where the bees enter the building. So do not just rely on thinking you know where it is. Another thing to remember is that if insulation gets in the way it will affect readings and you may need to look in other areas or use endoscopic cameras to search for the colony manually.

The Use of Endoscopes

The word "endoscopy" is derived from the Greek language, by combining the prefix "endo" meaning "within" and the verb "skopein", "to observe or view". Originally endoscopes were designed to look into the human body, and evidence in the ruins of Pompeii were found of instruments that were believed to be for this purpose.

In 1805 Philip Bozini made the first attempt to look inside the body with the Lechleiter. A light conductor made of two parts, a) the light container holding the optical device, b) the mechanical device holding the viewing instruments designed specifically for the organs that were to be observed.

It wouldn't be for another 79 years before the first rigid gastroscope would be available. Obviously, these had limited use as a rigid instrument. In 1932 Dr Rudolph Schindler invented a gastroscope that could observe the insides of the stomach at a slight bend. Further developments were made over a number of years in the medical industry, and the invention of a rod-lens optical system by Harold H. Hopkins, PhD, in 1959 and the addition of fibre optic light transmission by Karl Storz in 1960 marked a breakthrough in modern endoscopy.

In 1974 a number of European architects began research into the practical applications of endoscopes in urban planning. Which resulted in the many varied jobs they are used for today, including drain survey work and internal structure assessments.

The modern-day endoscopes are a must have piece of equipment in the bee removers surveying kit bag and are available in so many forms. The little handheld ones are really useful for small close up cavity work, but for the out of reach areas, a longer Bluetooth operated camera is more practical. Particularly where chimneys, and wall cavities are concerned.

There are many makes and specifications for endoscopes. Simply look to purchase the one that is most suited to your needs and budget. Photo by Clive A Stewart

As with any equipment there are cheap cameras, and there are expensive cameras. Be sure to get something that provides good image quality. As this helps provide better reports for the client. Never has the phrase "a picture says a thousand words", meant as much as it does on a client's report.

The pictures back up your summary of what is going on and how you will approach resolving the issue for the client. They help the operative to locate the parameters of the colony. This will assist in gauging the colony's size, and on occasion when a thermal imaging camera has failed, help locate the colony.

They may also be able to ascertain the colony's health though this does depend a lot on how the bees are viewed, and what the observer can see of the comb. This is where training and education in the use of the equipment comes into play because operating an instrument is only half of the task, understanding what you are seeing and evaluating that, and creating a format that the customer can understand is equally as important.

Insertion of an endoscope camera into a ceiling void in where a thermal imaging camera has been unable to pick up a heat signature, in order to determine where a colony is can be quite useful.
Photograph by Clive A Stewart

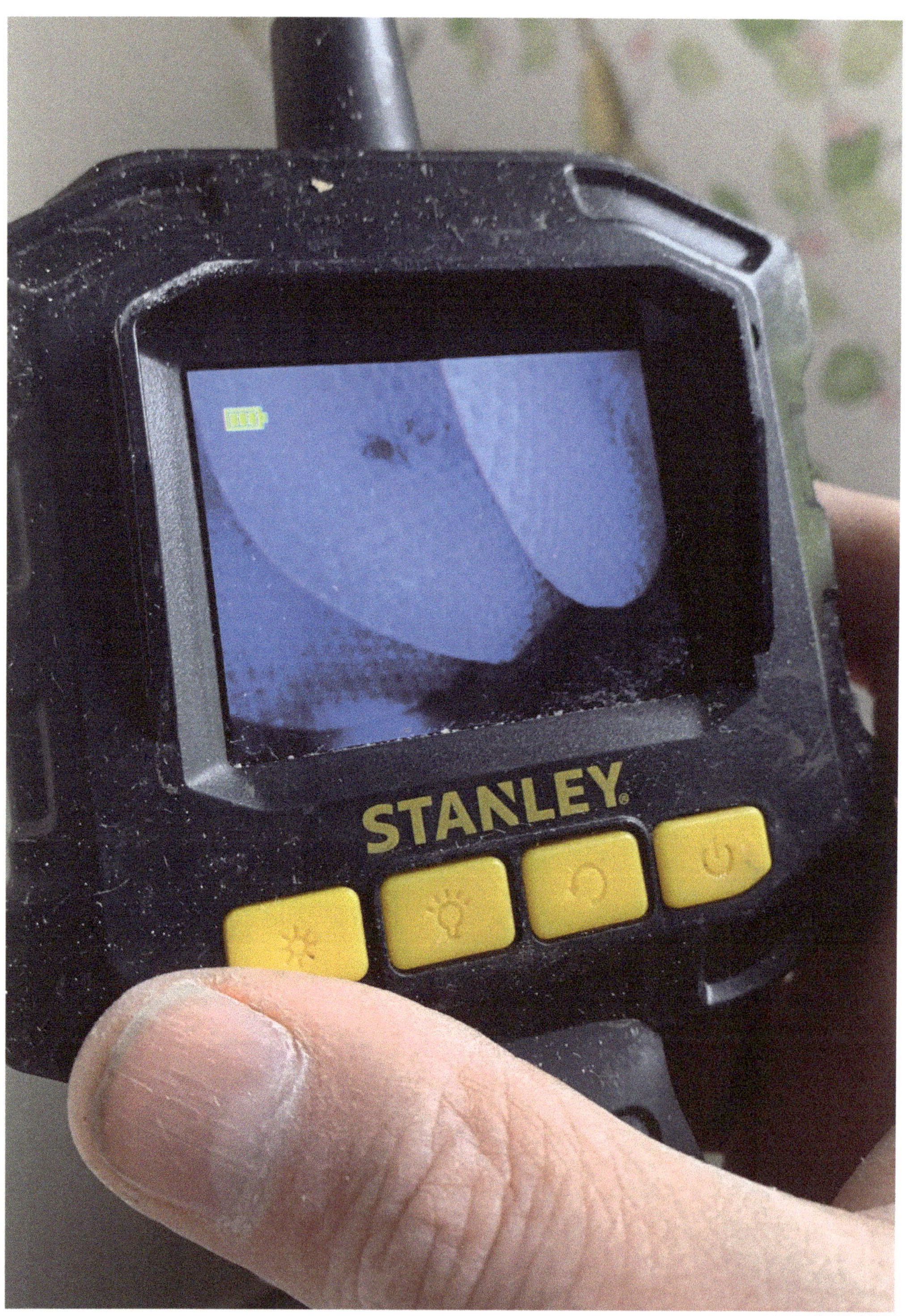

Typical image of what can be seen using an endoscope - photograph by Clive A Stewart

Part 4

Colony Removal

Part 4- Colony Removal

Reasons for removal

In general, the first and main objective for removing a feral honey bee colony is usually the health and safety implications of the local environment that the offending colony is located in, and the occupants of that building, and its surroundings. These can vary from public buildings, such as schools or hospitals, to commercial office blocks, industrial factory settings and domestic properties.

The one thing all these situations have in common is people, and as already stated in this book, an estimated 5% of the population suffer severe reactions to bee stings. Whilst understanding that bees will not generally go out of their way to sting someone unless provoked, it is considered that they may become a potential public health issue.

However, there are instances where colonies do not need to be removed and can be left to get on with doing their thing. In fact, there are possibly as many colonies left as there are removed. I know of one instance in Cambridgeshire where a feral bee colony is situated in the wall of a public house, so named the "Bees in the wall". It is a local tourist attraction and was aptly renamed after the bees moved in around the 1950's. It's also understood that it is the only pub in the world with this name.

Photo of the Bees in the Wall in Cambridgeshire. Photograph by Clive Stewart

Each situation will require a level of evaluation and consideration needs to be given to those colonies that maybe in situations by where they are in constant sunlight, and where honey may potentially fall into the property as this can and will create disastrous results internally if ill advised. Flat roofs and chimneys are a prime example.

I have seen in both these situations the results of poor advice. The covering of flat roofs come under extreme heat build-up in the summer sun, as a consequence of this comb containing honey will fall and collapse on to the floor of the ceiling below. This inevitably ends with honey dripping through cracks and or even light fittings.

Collapsed comb of an exposed colony that had been dripping honey through the ceiling and window frame of the office below. Note how the comb has no regimental structure usually found, and how it has all fallen to the floor of the exposed void. Photograph by Clive A Stewart

The window frame below the colony in the previous photograph.
Note the honey running down the wall. Photograph by Clive A. Stewart

The consequences of honey comb falling down chimneys is common place in favourable weather resulting in a strong nectar flow. The comb often becomes heavy, and the wax not strong enough to hold the weight of the abundant harvest coming in. I have seen numerous carpets ruined in these countless situations.

Comb and bees fallen down in to a disused fire place in the heat of summer looking a little bedraggled. Luckily the client only had small quantities of honey come down too.
Photo by Clive A Stewart

The bees can however, live out their life cycles quite happily without the interference of man, some will naturally die out, only for the combs to be recolonised at a later date. This is not generally down to any one particular reason, but possibly an indication of many accumulative factors. Weather, the lack of food stores, disease, queen condition, and parasite levels, will all play a factor in how well a colony will do during the winter months.

When bees are trapped inside a void they will always look for the easiest way out. These bees had found their way out through the skirting boards of a client's bedroom who had attempted to open the painted shut windows. Photo by Clive A Stewart

I regularly hear the poor advice that colonies cannot survive through the winter and the occupant should seal up the entrance to prevent further issues when the bees have quietened down in late autumn, or early winter. I am then, more often than not, contacted in the following spring by those very same, but alarmed people that took this advice. Asking, or pleading with me to come and remove the bees from their kitchens or bathrooms, as the 'believed to be entombed' bees have discovered another way out.

It's not uncommon for bees to get in the way of home improvements, and modernisation. One of the most popular additions in recent years has been to install a wood burner in homes with established chimney systems.

A colony in a chimney – photograph by Clive A. Stewart

Chimneys are often a favourite area for honey bees to populate, and another instance where the correct advice is paramount. I hear countless times that bee experts have informed the occupants of their beautifully decorated homes that if they are able to, that they should light a small fire to encourage the newly arrived bees to move on. This is ***not*** a good idea, and a number of things need to be considered here.

Firstly, it is most likely that the occupant has only recently become aware of the bees after they have arrived, and there is a potential for there to be some wax in the chimney, even after a few hours. One must remember that wax melts at 64°C, and the flash point is 204°C. It is also worth considering that everyone's idea of a small fire is different. The consequences of wax melting and falling down the chimney to then possibly catch fire is a very high risk and could have devastating results.

Secondly, if the said colony is large enough to have reduced the flue of the stack and prevent the smoke from being drawn up the chimney, I have personally seen this happen. It can blow back into the beautifully decorated home with some truly horrendous results.

Trained bee removers will have a number of tricks that have a much less damaging outcome to the situation. It will of course all depend on how long the colony has been in place. If the colony has had more than 24hrs to settle then it will be more luck than planned eviction, that the colony decides it isn't the home for them.

This will be due to the fact that by this point there will be the potential for eggs to have already been laid in the newly built cells. It then simply becomes a full removal, and the sooner it is carried out the better. The longer a colony is left, the further down the chimney it will build. This is usually around a foot per year in your average chimney. Consequently, this would result in more of the chimney being opened.

It's also worth noting that your enquirer may have only recently moved in that winter, and the surveyor obviously hadn't noticed any bee activity at the time of his visit in the house buying process, and this is the first time for the new occupants to have met and experienced their resident squatters.

Bees seen from below the colony using a chimney camera to help determine how far down the colony is situated. This helps in any potential scaffolding design requirements.
Photo by Clive A Stewart

Not a particularly deep colony shown from the crown of the chimney stack.
Colonies have been known to be as deep as 2 or 3 metres down. Photo by Clive A Stewart

Roof repairs are another common area for renovation or modification, and inevitably has bees coming into contact with tradesmen and the occupants of the building. Flat roofs in particular can be troublesome. This is often because the offending roof has possibly had a lazy builder come along and simply over laid the roof and in doing so has created a second void space. Which whilst is fairly small in space is actually ideal for colonies to set up home in.

For roofs, it usually takes up a lot of resources to locate the colony as thermal imaging will not always pick up the colony, and so other methods will need to be employed. Further problems can be encountered when the overlaid roof joists have been laid in the opposite direction to the original roof. Even more so if it is decided that internal removal is sought.

Types of Removal

I'm confident that you will now be aware of the fact that if there is space in a cavity of a wall, floor, ceiling, or chimney, honey bees have the potential to make it their home. It is of course the decision of the owners of that building as to whether or not they stay.

Having decided that the bees will need to be relocated, a professional bee remover should be sought. I must stress again that a survey of the offending honey bee colony, and its location are paramount to a successful removal operation. Failure to carry out this procedure will likely cause some difficulties and complications during any removal, with some possibly disastrous and catastrophic results.

There are mainly two forms of removal, these are 'Cut-Outs' and 'Trap-Outs'. We will look at both closely, and outline the advantages, and disadvantages for both the client, and the bee remover themselves. On the completion of a detailed survey, the bee remover should be able to advise their client of the best and most economical way in removing and relocating the honey bees.

Cut-Out Removal

As the term implies, a Cut-Out is exactly that. The comb, and attached bees are exposed, and cut away from their position within a cavity. Removing the comb and bees together gives the colony the optimum chance of survival. It also allows the bee remover to check the colony's health, and to examine comb for notifiable diseases.

This method also goes a long way in preventing secondary pest infestations, such as larder beetles, moths, flies, and rodents. Squirrels and mice are very partial to honey on the comb. There are optimum times of the year that this can be carried out to maximise the colony's survival rate. However, it is not by any means constrained to these times and can be utilised all year round.

Be under no illusion that this method is destructive, and those carrying out the removal must have a good, and clear understanding of building construction. All care must be taken regardless of where the colony is, and attention must be given to service supplies, building infrastructure, and the safety of the building's occupants. These along with other safety implications, which will be building specific, and must be evident in any work plan documentation.

Bee removers must note that all work plans must be strictly adhered to. Any deviation from the plan must be authorised by the client or their respective health and safety officer.

There are within the Cut-Out removal works two ways a colony can be exposed. Below is an overview of these, along with the pros and cons for each.

Equipment for Cut Outs

The amount of equipment required for a cut out is huge and will vary from job to job. What is needed to cut a hole in a ceiling will not work on the removal of bricks from a chimney, but on the whole, I try to carry everything that I'm likely to need from a screwdriver to specialised equipment like the 'Bee Vac', and I have tried to cover everything in the tables below.

PPE	Comment
Bee Suit	Ideally a suit with a kite safety mark, should be identified in your RAMS, and site-specific plan of works.
Gloves	Domestic rubber gloves
Safety Footwear	Steel toe-capped boots
Dust Mask	Where necessary be aware of room under the veil
Ear protection	Where necessary use sponge style to fit under veil
Bump cap	Have found peaked bump cap most suitable for building site situations

General Use	Comment
Hoover	Worth purchasing suitable make where spares are readily available
Bee Vac	Light and easy to clean
Short and Long Hive Tools,	Extremely useful to assist in cutting out comb. Paint scrapers can also be used for this job.
Smoker	Should only be used if your risk assessment allows and you are covered by your public liability insurance to do so. Cardboard is a good fuel that has minimal embers when used correctly, and with a spark guard fitted. It must never be left unsupervised.
Water Atomiser	Suitable alternative to smoker but has limitations. Assists in providing water and keeping the colony cool. It also has the effect of preventing the bees from flying when they are wet.
Electronic Smoker	An alternative to the traditional smoker and works in a similar way to a vape providing an ultra-low heated smoke with no embers.
Repellent	To deter bees moving into areas in which they cannot be reached from the current access point.
Ladders, Steps, & Step ups	Ladders are only for access to working platforms or short period of light work.
SDS Drill, Driver, Multi Tool, Circular Saw, Disc Cutter	It is worth noting that all these power tools listed will inevitably become covered in honey. An expensive make will become as quickly gunged-up as a cheap one. Honey has the ability to ruin electrical equipment corded or cordless very quickly.
Extension Lead	For corded electrical equipment
Hand Tools, Hammers, Bolster Chisels, Crow Bars, Screwdrivers, Spanners, Etc..	A broad selection of hand tools for the use of opening up is handy to have, as things are not always as they seem, and you may need something out of the ordinary to assist.
Polystyrene Nucleus or Hive	For re housing comb and colony. Tip: Use a queen excluder under brood box if using a full hive. This will help to prevent the bees absconding from the hive once they have been rehoused.
Hive Frames or Comb Clamps	For securing combs inside chosen hive or nucleus box (nuc)

Zipped collapsible laundry bag	Useful for transfer of bees from bee vac if hive is likely to become too full. This allows bees to get enough air to keep cool.
Loft Insulation, Foam	To assist in proofing where applicable. Must not be used in walls or chimneys.
Bucket, Clothes, Dust Pan & Brush, Bin Bags, Wire Brush	Keeping a site clean is very important and saves time at the end of the day.
Washing Up Liquid	Ideal for assisting in cleaning up honey spills and will work in both cold and hot water.

External Use	Comments
Mobile Scaffold Tower	Useful for those easy to reach colonies that haven't been established very long. PASMA certificate required for professional use. (Prefabricated Access Suppliers and Manufacturers Association)
Drop Board	Specifically used in chimney removals to prevent comb from falling down the chimney.

Internal Use	Comments
Polythene	Min 1000 gauge for covering the floor on internal removals.
Duct Tape	To secure, and or seal polythene sheets in place

The van packed with all the necessary equipment for a typical day in bee removal.
Photo by Clive A Stewart

Bee Vacs

Whilst the majority of equipment is easily found in most DIY stores or builder's merchants, bee vacuums are a little specialist. Those available in the UK work on the same principles as saw dust collectors do in a wood work shop. Effectively a vortex cylinder or box that collects the bees before being sucked into the vacuum cleaner itself.

There are some strong and well-made wooden ones available, that work very well. These have various adaptations to assist with the task in hand, and to reduce damage to the bees. A sloped floor at the entrance point, sealed lids with securing clamps to prevent the box from coming apart, sliding floor to enable the operator to transfer his bees to a nuc or hive without the need for second manipulation. The disadvantages to these are that they tend to be heavy and not very easy to clean & sterilise.

Then there are the plastic bucket types available too. Lighter and easier to clean makes these types of units more manageable when working on high access platforms. These are also not without their disadvantages as they usually require a second manipulation. Whilst this may not necessarily be a major issue it can sometimes be a little annoying. Some have a catchment insert which is very useful if the vortex gets a little full, as this can be removed, and the bees allowed to cool whilst being contained and replaced with another allowing the bee remover to continue. Both have manual flow rate adjustment and smooth bored hoses to prevent damage to bees when being vacuumed up.

Plastic bee vac in use. Photo by Clive A Stewart

Wooden bee vac in use to the right of the picture.
Photo by Marcus Collings (Lakeside bee services).

All these vortex type vacuums use a controlled vent which allows the operator the ability to adjust the suction flow rate of their chosen vac. Using a bee vac has a level of skill. If it is used with the vent shut too tightly and the bees sound like a hail storm it is likely to increase the mortality rate within in the vortex, too slack and the bees won't be sucked up.

I was once told by one of the producers of these units that anyone could use these, and that may be true. However, it's not as straightforward as it seems, and it's not a bad idea to go and have a little training on these pieces of equipment. Those that do produce them usually do some level of training in their operation, which is often free.

Another point worth noting is that whilst these are a very useful pieces of equipment in the armoury of the bee remover, the operator should give consideration to removing as many bees as possible manually on the comb. This not only helps reduce stress but assists in alleviating mortality rates in the vortex. Bees can soon get hot and sweaty inside, and they should not be left in the vortex for long periods of time. The techniques that I employ on chimney removals today rarely require the use of a bee vac at all.

Repellents

Bee repellents have long been used in the production of honey by beekeepers that have wanted to evict bees from honey supers. Usually spraying an absorbent pad with a concentrate of some formula that the bees dislike. It is invaluable in the world of bee removal in preventing bees disappearing into cavities out of reach or loft spaces during roof removals. There are two that are commonly used Bee-Quick and Apifuge. They can also be used on areas that are being scouted out by a nearby swarm looking for a new home.

Apifuge is an aerosol made from a number of organic essential oils, and smells a lot like thyme. It is supposed to have a calming effect and is recommended to put on your hands prior to any manipulation in beekeeping operations. However, when using it liberally in the bee removal environment it seems to have an adverse reaction and certainly makes them move out the way but definitely not in a calm manner. In fact, quite the opposite in my experience, although it certainly does the job.

Bee-Quick is made from diluted benzaldehyde and smells a lot like almonds. It is a strong acid and needs to be used with extreme care as it can cause irritation if you get it in your eyes or on your skin. Bee-Quick does not arouse or aggravate the bees quite as much as Apifuge does but will work quickly in dispersing them.

Some people also recommend Vicks VapoRub, and it does have some effect on them. However, in my experience it is slow to work, and you need lots of it. It is very useful to deter those colonies that are just scouting for a new place to live, and have not moved in yet.

A selection of repellents –
photograph by Clive A Stewart

Framing

As you cut the comb from the area in which you are working, generally speaking, it is a good idea to organise some of the comb into a hive or nuc box so that when back at the apiary site it will be easier to manage. There are two ways that this can be done.

One way is to use the frames of your chosen hive design and use elastic bands to secure in place the pieces of comb. Two or three elastic bands is more than sufficient. Though I have seen some used several in all directions. This makes it both difficult for you and the bees.

To help support your comb whilst doing this you may choose, as I do, to wire the frame, and this can be quite helpful, and act as a third hand, as it is not easy to do when they are covered in bees.

You can of course shake the bees off but remember to do this over your hive or nuc box, if you choose to do so. As time goes by you will often see the bees throwing out the elastic bands as they have chewed through them and decided that they no longer need them.

Elastic bands in use for framing of comb.
These will be chewed off by the bees as they become established. Photo by Clive A Stewart

Another option is the comb clamp. Two bits of wood simply clamped together over the top edge of the removed comb using plastic or metal frame spacers to secure them together. It can be a little fiddly to master but once learnt it is so much quicker and easier than securing elastic bands to frames, and there is little disturbance of bees too.

Using comb clamps has the advantage of not having to manipulate elastic bands around a large frame and bees. The disadvantage is that they will brace to the sides of their new home. - photographs by Clive A. Stewart

This is how recovered combs fit into a standard nuc- photograph by Clive A Stewart

Of course, it should go without saying both of these operations are best carried out over your chosen transportation receptacle, be it a hive or nuc box. Just in case the queen happens to be on the comb you are working with. The last thing you need is to see a cluster of bees on the grass below the scaffold.

Bees gathering around a dropped queen found part way through the removal.
By Marcus Collings (Lakeside Bee Services)

Internal Cut-Out

Internal removals are probably one of the safest ways to remove a colony, from the public's point of view, as the majority of the guard bees at the beginning of a removal will be contained within the building. Though, it does not come without its share of complications.

Services such as gas, water, and electricity, often run through areas in which we wish to expose, and these must be searched for either during the survey, or on the day of removal prior to cutting into any building material that the colony is situated behind.

It's also worth remembering that any buildings built before and including 2005 should be considered in regards for the provision of asbestos and any current legislation adhered to as a result and highlights the need for asbestos awareness training.

Internal works are very messy and some more than others. Be aware that seasonal timing can assist with reducing this. If an internal Cut-Out is sought in July, consider putting off till the following Spring. This will help with both numbers of bees and the amount of honey that you have to deal with during the operation. The stores that were in situ in July will have been dramatically reduced by the overwintering colony.

Preparation is paramount when carrying out an internal removal. It is good practice to have the room cleared of all soft furnishings, valuables, and furniture where practical. It is advisable to inform your client that re-decoration maybe required on completion of the removal. This can be, for some removers, an extra form of revenue from the project.

Cover everything left behind with polythene sheeting. Do not use cloth sheets as honey will run through these. Floors should be covered with a polythene at a minimum of 1000 gauge as this will be generally strong enough not to tear during your working activity. Be sure to tape this all the way along the skirting board, or more favourably part way up the wall creating a temporary bunding. Avoid spot taping, as this will inevitably leave gaps by where bees, dirt, and honey that will fall down behind.

The contents of the bees' guts are hard to get out of a cream carpet. If a room is large, it can be sometimes easier to partition it off using plasterboard props to trap polythene sheets up to create a temporary barrier. It's a good idea to use black polythene for this and allow the bees to naturally be drawn to the light of the windows. Refrain from opening windows as this will create chaos outside and hinder the clearing of bees. Use your chosen bee vac to collect these intermittently.

Left, the use of polyethylene in an internal removal is paramount, and it is recommended that the floor cover should be a minimum. Photo by Clive A Stewart

Internal Advantages	**Internal Disadvantages**
Reduced number of bees externally, in turn reducing risk to passing public.	Messy and usually involves redecoration, particularly if ceilings, or studded walls need to be opened up. Honey Gets Everywhere.
Reduces the requirement for extensive scaffolding. Although, there is sometimes need for scaffolding for high ceilings.	Can sometimes be working in confined spaces.

Tip: Honey inevitably falls to the floor and, as much as you try, you will not prevent it from getting it on the bottom of your footwear. Work with a spare pair of shoes, leaving them at the edge of the work area to change into and out of when moving away from the work area. Even if you have sheeted your entire path in and out of the property. This helps reduce the spread of honey around the house when you realise you have forgotten an item in the van, because you will.

Tip: Where you have rooms with more than one window, it is a good idea to blackout all but one window. I prefer the one closest to the work area, in order to keep the work space as tidy as possible. This will assist in reducing the need to clean all the windows and give the bees one area to congregate in.

Internal cut-out in progress – photograph by Clive A Stewart

External Cut Out

It has to be said that I much prefer external removals, simply due to the comfort of the open space. However, this type of removal does not come without its own set of complications. The obvious one being the general public. I understand that it is not every day that you get to see some loon in a funny, bright white suit, looking like he's enjoying himself on an industrially erected stage. I have on occasion understood exactly how Elvis, and the Beatles felt, and whilst my audience have not been screaming or throwing their underwear at me, they have from time to time got a little bit too close for photos, regardless of the warning signage in use, and barrier tape strewn everywhere like it's some sort of royal celebration.

Dependent upon how busy an area is I will increase safety measures to incorporate crowd control measures. This could include a pedestrian diversion to the opposite side of the road, or enclosing workspace in a commercial operation. It should be noted that any obstruction of the highways and byways in order to carry out work has to be applied for and approval granted before any works can go ahead.

External works often mean that building materials need to be removed, and more often than not this is done at height. Safe access to these areas is paramount, and I cannot emphasise enough that this type of work should never be attempted off a ladder. As has already been stated in the health and Safety section if this book, scaffold is by far the safest way to access any colonies at height. You should consider how much space you will require, and it is definitely worth asking your scaffolder for a pulley, particularly when there are three or four levels to the top working platform.

External Advantages	External Disadvantages
No internal damage or disruption and does not require redecoration.	Access equipment such as scaffold or Powered work platforms are required.
No requirement for covering any areas, as foraging bees and wasps will clean any split honey.	Public safety is a heightened risk.
	Building fabric is required to be opened up, and reinstated.

Tip: Many hands make light work. Try to work in pairs, this not only increases safety, but also reduces the workload on one person at the end of a hard day.

Tip: Keep the area clean at all times, regular washing up of spilt honey and wiping hands not only reduces cleaning time at the end but will also reduce the likelihood of honey bees robbing, which can become problematic at certain times of the year.

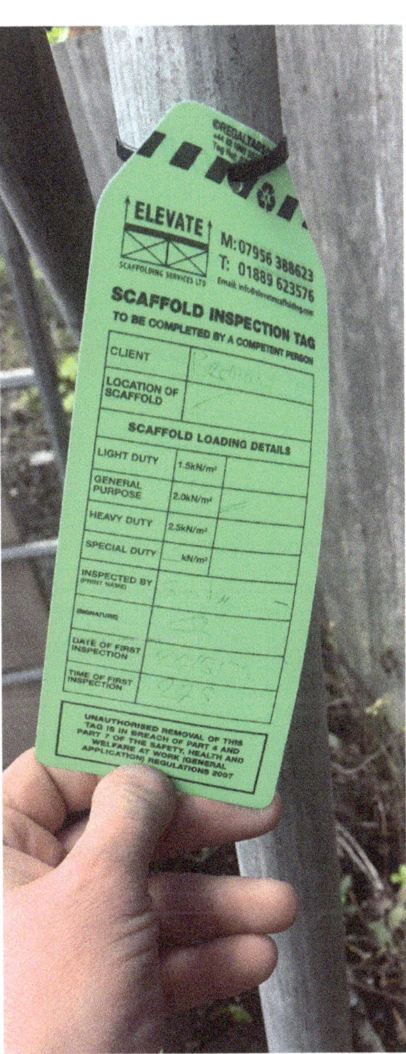

Scaffolding should see safety features such as toe boards, safety gates and where necessary ladder locks. Before getting on any scaffolding, it should be noted that it must have a 'scaff tag' or you may find that your insurances will be void if anything should go wrong.
Photographs by Clive A Stewart

Methods & Techniques

Many techniques, and methods are used throughout the process of a Cut-Out removal. Not only will these techniques, and methods differ from site to site, and situation to situation, but they will also change between individuals.

Some removers will prefer to vac first, and then cut away the comb that is free of bees, choosing to reunite both later in the operation, either on or off site. Whereas others, like myself will prefer to vac up at the end choosing to keep as many bees on the comb as is physically possible. There is really no right or wrong way in completing a Cut-Out operation. However, I should perhaps explain why I choose to vac last wherever it is possible.

The process of removing a colony is quite stressful on the bees, and none more so than when being sucked up by a vacuum. Whilst there are many design features in most bee vacuum setups that prevent the damaging of bees during this process there are inevitably going to be a number that will not make it. There is always a level of mortality during cut out removals, and it is in the interest of the bee remover to reduce this by as much as possible, in order for the colony to have the best chance of survival.

Another action to reduce stress in a removal is to remove a colony in Spring rather than at the height of summer because of the smaller colony size. A colony in Spring is less than a third of the size it is likely to be in the height of the season and will therefore be able to cope with stress much easier than it would if it were to be taken out in summer when numbers would cause stress to linger for some time. To assist in reducing mortality rates, I attempt to take apart the colony in a similar way as you would manipulate a top bar hive. Carefully cutting out the comb, examining it for disease, stores, and making sure that brood in all stages are present, I then suspend the comb either in frames or by using comb bars. The bees that are removed on the comb are under a lot less stress than those sucked up into the bee vac.

Bees on the comb being prepared to be framed. Photo by Clive A Stewart.

Bees on comb with brood at all stages prior to being framed. Photo by Clive A Stewart

Remembering that you have completed a thorough survey and done the necessary checks and tests for asbestos in those buildings falling under the necessary remits, and you have gained the appropriate permissions from your local authority conservation officer in regard to listed buildings where required. It will be necessary to expose the comb for removal. Some will be very easy, and others not so, a very small number will be impossible, and at this point you may consider a trap out.

However, it is immensely important that you understand fully the situation that you are about to take on, and those safety implications involved. When exposing a colony, it is best to try and expose as much of the comb as is safely possible. In floors, ceilings, and some roof situations this is quite simple. However, where bees have built in cavity walls, and chimneys this can be extremely difficult, and if the remover is not experienced in this line of work, then they should consider finding someone who is, and look to assist them in these works.

Removals from floors and ceilings.

Theses removals, whilst potentially in similar location, as internal removals, are a million miles apart in their approach. Floors are relatively simple and straightforward. In fact, it is almost like dismantling a top bar hive, only without the bars. Honey and bees can be lifted out with very little effort and in most instances with very little use of the bee vac.

Exposing a colony via a floor is not without its risks, and you need to be aware of services running underneath the floor material, such as water, or gas pipes, and electricity. There are many devices on the market, and these are relatively cheap in comparison to the cost of repair from damaging any of the afore mentioned services. Checking for these services should form part of your survey early on, and or carried out prior to making any necessary cuts.

Whilst ceilings are in a similar area, they are far more troublesome. The big difference of course is that you will be working from underneath. As a result, you can expect the honey to do as you would expect and react to the forces of gravity. In short, it will rain honey, and in the height of season this can be quite a lot. It is not unheard of for me to change suits two or three times on the more complicated and difficult ceiling removals.

Again, it is worth remembering that services need to be considered, and whilst most will run just under the floor I have seen and known of situations where cables lay on the floor of the ceiling. It is so easy to check, with the right tools, that it should become second nature. Caution has to be made to those ceilings with textured decorative coatings as early manufacture of this product contained asbestos. Any textured coating in buildings built in 2005 or older and you should have the area tested and consider bringing in a specialist company for that removal.

Bee colony under a floor. Photo by Clive A Stewart

Bee colony situated in ceiling void. This will have been under a flat roof. Note the sponges in two inspection holes to prevent bees wandering into the vacant voids. The inspection holes were made to determine whether or not there was any over flow comb. Photo by Clive A Stewart

Removals from Roofs.

The removal from this area is often fraught with the complexities of working at height. Opening the colony is generally easy to execute, unless you have Collyweston roof tiles or a thatched roof. Both of these will require the assistance of specialists in order to enable you to carry out the works. When carrying out a survey of the roof area, be aware of ACMs (Asbestos Containing Materials) in the soffit, old felt, and some roof tiles.

On completion of a thorough and successful survey it is advisable, where possible, to go from above. This is because the comb is exposed more evidently and there is least resistance on extracting the fins than when having to work from underneath scraping against the material on which they are attached to.

However, from time to time it may just be as easy, if not easier to remove the colony from inside. It can potentially be cheaper too, though this is dependent on a number of factors, and not necessarily to be relied upon. It is definitely much easier in the earlier months of the removal season when bee numbers, and stores are low, and you are not rained on from above with constant flow of honey in the same way as you would for a ceiling. (See preparation for internal works).

Exposed colony in roof space. Note the white powder at the bottom left hand corner of the exposed bees. This colony had previously been treated and would require extra care during removal. Photo by Clive A Stewart

An exposed colony in a dormer roof space. These spaces are of particular interest to the honey bees, and it is not unheard of for them to fill these voids to the brim.
Photographs by Clive A Stewart

Removals from cavity walls

Cavity wall removals are difficult to carry out due to the required knowledge needed to be able to open them up without bringing down the house. Simply put, if you have no building knowledge or experience in this trade you must not attempt the removal of any brickwork without seeking the assistance of a professional.

Whether or not you are carrying out the reinstatement works there is a need to gain as much information about the colony as is possible. How long it has been in situ, and how big the colony is will be critical in working out exactly what is going to be needed before removing any brickwork.

The age of the building is important in regard to the potential for the presence of asbestos. Is there any steal work internally? Are there any pipes or services present in the work area?

Careful consideration for the brick work is required when opening up walls as not to make too bigger hole that it has profound detriment to the structural integrity. If you are not competent in executing such work you must employ the services of a professional. Under no circumstances should you attempt a removal from a cavity wall without training. Photo by Clive A Stewart

Colonies can get to be quite big when conditions are favourable. Photo by Clive A Stewart

Removals from chimneys

Chimneys have the combined issues of both walls and roofs, by where they are constructed in brick or stone and are predominantly at height. They also vary in design and construction massively, both from house to house and building to building.

They all have the ability to be very problematic, and no one will be the same. Dealing with the colonies within side of them depends a lot on their current use. If a chimney has a flue situated in it, most of the time, but not always, a colony can be found to be situated between the outer wall of the flue, and the inner wall of the original stack. The flue has the potential to be asbestos, as does the lining of the stack. The discovery of asbestos in a chimney is rare, but none the less very much worth mentioning, as clean-up costs for asbestos often run into eye watering amounts. Without potential fines from HSE too. This should obviously be considered prior to any works that may require the dismantling of a chimney in order to remove the colony.

Again, I cannot stress enough the importance for independent asbestos awareness training, and this should be sought as a minimum requirement when looking to carry out any bee removal works.

One thing that all chimneys require is a thorough survey and not just of the inside but the outside too.

- ▶ Does the chimney lean?

- ▶ Has it got plants growing out the side of it?

- ▶ Any obvious degradation to the mortar or bricks?

- ▶ Does it have an ariel?

- ▶ Can it be easily surveyed internally, or will it require vents or bricks removing?

All this information is required and more to assist in creating a plan of action to remove the colony.

There are a number of chimney cameras available, and I have in the past used my phone camera attached to a cheap set of drain rods when the chimney has had no bend as an emergency option to get me out of a tight spot. These help to take a look inside and observe exactly how far down the stack the bees have come.

Generally speaking, it is estimated that bees will build as much as 300mm (12 inches) in a year down a chimney. Don't forget to add a year to your client's recollection of when they noticed them, as it is likely that they were not seen for the first year.

All things considered this is just a very brief over view of colonies in chimneys and it is strongly recommended that they are not undertaken as a newcomer to the industry without the appropriate training.

Bees love chimneys with this type of "pepper pot" cowl on them.
They make excellent homes away from predators and the damp ground. Photo by Clive A Stewart

Removals from Trees

The removal of a colonies from trees is not a common one and only tends to come about due to the colony becoming exposed due to storm damage or there is a need for the tree to be taken down, and this is usually the first time the owner will have been made aware of it.

If a colony has become exposed in a tree that has suffered storm damage it is usually quite easy to deal with as all the comb is relatively accessible. The only possible complication will be the surroundings. Be sure that it is safe to carry out the necessary work without injuring yourself or those working with you and be mindful of any public access.

A colony situated in a tree labelled for felling has two options. It can be either trapped out or simply removed to enable cut out work to be carried out off site in a more secluded and safe area. As to which one you can use will be dependent on the time available in which to carry out the task. Trap-Outs are notoriously time consuming, and therefore not suitable if the tree is to be taken down quickly. The alternative is to seal up the colony and cut below and above the cavity and have it craned on to a pick up or suitable vehicle to relocate elsewhere. Remember that this must be at least 3 miles away from the site or the bees will return to the original site. Sealing up can be done simply using sponge to fill the hole. Suitable adhesive tape can be used as further securing of the sponge.

Once it is delivered to the chosen site it should be put back up in its original orientation until it is extracted. Again, it can be trapped out or look at splitting the section of tree open to expose the colony.

Bees in trees are not usually an issue until either the tree is needed to be moved, or as in this case they have been exposed to the elements due to some recent bad weather.
Photo by Clive A Stewart

Dealing with such situations is fairly straightforward in storm damaged trees but can be a little trickier when the comb is deep inside. Tree surgeons are often required along with specialist lifting equipment. Trap outs are also suitable for this to, though will be time reliant.
Photo by Clive Stewart

Trap-Out Removal

A Trap-Out, is simply a method by which bees are allowed to exit the colony entrance but are prevented from re-entering. This is usually done by the means of a wire gauze cone. I have seen some beekeepers in the past try using porter bee escapes that are used to empty honey supers on managed colonies. This has had limited success, due in part to the misunderstanding of the colony's situation, and the utilization of the equipment used.

The use of a surrogate colony assists in providing the returning foragers with somewhere else to relieve them of their gatherings. The big positive for this method is the lack of destruction involved in evicting the colony. However, it has many downsides to its success. The inability to recognise any brood disease is a huge disadvantage.

The requirement to revisit on a regular basis can be disruptive to both your own plans, and that of your client. It can only be carried out when the bees are flying and therefore limits this type of removal to Spring and Summer.

House bees are soon deprived of food coming in and are resigned to use the stores in place to feed the young larvae. If these are deprived of stores, then the obvious starving would ensue. It is unknown whether the queens leave on every occasion or not. I have never witnessed them doing so myself. Reports from those who claim to have performed Trap-Outs, speak of small clusters being found in nearby vegetation or even wandering into the provided surrogate colony.

Due to the many disadvantages, I only ever use it in extreme situations where access to the building is physically impossible. Whilst this option looks potentially favourable due to its lack of destruction to a property. It does not operate in a favourable way for long established colonies, it can however be very useful on those colonies that have truly only just moved in, where access is easy enough to enable it to be installed without hindrance.

Note, how the bees at the base of the cone are trying to orientate to the original entrance.
Photo by Clive A Stewart

A very heath Robinson style trap out, but space was limited, and the situation dictated the inability to be able to fix to the wall. However, this does show the flexibility in the trap out method.
Photo by Clive A Stewart

Method

On deciding that a Trap-Out is the best option for the colony in question, logistics, and positioning become important factors for a successful outcome. You need to place your surrogate colony as close to the current entrance as is physically possible in order for the lost foragers to easily find somewhere to deposit their supplies on deciding they can no longer find their way back in.

Enough space needs to be allowed for the install of the wire gauze cone. This should be secured over the entrance with the exit of the cone protruding past the entrance of the surrogate colony by around 50-75mm. This can be done in many ways, such as stapling or screwing the cone directly in place. What I do is fix the cone to a face plate, and then secure the face plate to the substrate be it brick, stone, concrete, or other such materials. Be prepared to block any potential entrances around the faceplate.

The returning bees are immediately disorientated and begin to look for alternative ways in. Believe me when I say they will go to extraordinary lengths to gain access to their original home. I have personally seen them gain entry over three meters from the original entrance, and this is why regular revisits are needed, as these will require blocking up in order for the process to have any chance of success.

Equipment for Trap-Outs

There are simply two main parts to trap out equipment. The Cone, and the Surrogate colony. The cone should be around 40-50 mm at the base to allow sufficient coverage of the entrance, no more than 9 mm at the exit to allow two bees to exit at a time, and around 275-300 mm in length to allow the 50-70 mm past the entrance of the surrogate colony. I do just give my cones a little bend upwards to move the exit out of the flight line of the entrance and deter the bees from potentially re-orientating back to the new exit.

It is favourable to use polystyrene nucs, for ease of carrying and application in most situations. It is a good idea to use those with the wide slit style entrance. I have found bees seem to be much slower in finding the small hole entrance used on most of the all-in-one units provided by the bee equipment suppliers.

The nuc should house two frames of brood in all stages with accompanying nurse bees. Do not put a queen in with these frames, as there is potential for the queen from the Trap-Out to emerge from the cavity and sense the presence of a queen in the supplied surrogate colony and will therefore abscond with the remaining followers.

It is important to have an array of suitable tools carried by your average handyman, for the purpose of fitting brackets and securing the described cones and face plates. Screwdrivers, wall plugs, drills, cable ties and straps, would complement such a tool kit, as well as upholstery foam to plug up any holes or gaps that the colony may find over the following days once the trap out is activated.

The tables below show the advantages and disadvantages of Cut-Outs and Trap-Outs

Cut-Out Advantages	Cut-Out Disadvantages
Colony diseases and abnormalities easily identified.	Generally, requires removal of building materials and creates large scale damage.
Not colony cycle specific and can be removed at any time of the year.	Has high levels of mortality compared with trap out.
Can usually be carried out with in the day.	In high season honey is a hindrance during removal.
Entire colony and comb is removed so less chance of other pests being attracted.	

Trap-Out Advantages	Trap-Out Disadvantages
Reduced damage to building, generally only requiring the drilling of holes to mount the bait hive on the wall.	Notoriously time-consuming average to be between 3 and 6 months.
Less disruption to the colony needing very little in the way of handling.	Colony cycle specific to be successful.
Almost zero mortality rate.	Can only be carried out when bees are active.
	Comb left behind attractant to other pests.
	Cannot identify colony diseases or abnormalities

Bee Welfare.

The intention of removing bees from buildings is not only to reduce the impact on our environment by not using chemicals, but there is also the importance of relocating a beneficial insect, that is not only vital to our own existence, but is as equally crucial to the survival of lots of other living organisms around us. It therefore should go without saying that the welfare of the bees, both during and post removal, is as important to the outcome, as the eviction is to the client that wants them removed.

Care During the Removal.

There are a number of things we need to consider during a removal and reduce where possible. Below are just some of the things that will assist in making the removal process just that little easier for the bees.

> ▶ **Stress:** Is possibly the one biggest killer in honey bee removals, and potentially equates to at least 50% of failed removals. An accumulation of one or more of the situations listed below will contribute to stress in the colony.
>
> **Prevention:** Follow good welfare practices, by providing a clean, spacious environment, in well vented transportation vessels. Be those open meshed floors, or zipped laundry bags. Supply water to assist in keeping them cool. This can be done using a water atomiser and wetting them down where necessary.

> ▶ **Squashing of Larvae:** During a Cut-Out large developing larvae may get squashed or damaged. This is a problem that the bees will need to clean up. Any liquids will be ingested, and full bodies will be removed by the bees. This has the potential to cause dysentery if there is an excess volume of fluids to clean up.
>
> **Prevention:** Refrain from cutting or squashing brood when removing comb. Clean combs of excess squashed fluids where necessary. Keep as clean and free of debris as is possible.

> ▶ **Honey Spillage:** Keep honey spillage to a minimum, this not only reduces robbing and clean up time on site, but also reduces the requirement for the bees to clean too. The colony has a lot to contend with, without the need to carryout extra cleaning duties. Excessive amounts of honey over the comb may result in bees being overwhelmed and drowning in their own honey. This can happen when space in the void is constrained, and the remover has limited access to the void.
>
> **Prevention:** Do not allow honey to run over those combs which you choose to keep for framing or clamping up. Keep as clean and free of debris as is possible.

> ▶ **Vacuuming:** It is very easy to switch the vacuum on and forget about it. Its use should be kept to a minimum where possible. The air within the bee vac will increase in temperature as more and more bees are drawn into it. Just feel how warm the bees are inside when captured in the vacuum to get an idea. This is not the only set back, as

natural air flow when the vacuum is switched off is limited in some models, and bees will suffocate, sweat and die if not attended to promptly.

Prevention: Remove colonies as early as possible in the year, as these colonies will be low in number and are unlikely to require much if any vacuuming. Where colonies are likely to be large in number, keep vacuuming to a minimum by only vacuuming at the end of the comb removal stage. If vacuuming is used to reduce numbers to allow comb removal, empty the collected bees into a well vented receptacle. A large, zipped laundry bag is a great piece of kit to have handy in such circumstances.

Some thought has to be given to the queen status of the colony. Finding the queen is the sign of a successful removal but is not essential to completing the task. She will often be found in the dark trying to keep out of the way of the light.

Precautions should be made to prevent the queen from scurrying off to other areas of the building. This is particularly important when carrying out removals from walls, and roofs, as they can and do disappear further into the cavity or in the case of roofs, the client's loft, making the removal awfully difficult and unnecessarily complicated. There are many different ways this can be done, but generally it is by creating a physical barrier with material such as loft insulation packed around the edges of the comb providing you have exposed the entire colony.

The use of bee repellents is another way of deterring them from disappearing, these can be used on their own, or simply soak some absorbent material such as card board or an old rag and placing it in a difficult or hard to reach place, use a stick to assist in getting it to the desired area.

You may be unsuccessful in finding the queen, and it can be particularly difficult when you are concentrating on many other aspects of the work at the same time, and even more so if the removal is at the height of the season where numbers will be at their highest. However, providing you are carrying out the removal between the months of April, and June there is a chance that your colony will manage to create a new queen, should you manage to lose or kill the queen during the removal.

You must ensure that during the process of removal you frame up or clamp at least one or two sections of comb that contain eggs and larvae of less than three days old for the colony to be able to produce a new queen. As long as she mates successfully and returns to the hive all will be well. It is fair to say that those removals outside of the months of April and June, should be considered for amalgamation if they show signs of being 'queenless' post removal, providing they show no signs of disease.

Post removal Care.

After all the surveying, planning and hard work, you have now retrieved your bees from whatever void you may have been asked to remove them from, and kept them as clean, cool, and calm as you could have possibly done. You're likely to have either, relocated them to your own out apiary or passed them on to your local friendly beekeeper who will have done the same. These bees are going to be stressed, and dependent upon the time of year, how big the colony was, and how easy the removal was will depend on how stressed they have become.

At this point there is absolutely nothing that can be done other than to pretty much let them sort themselves out. If you have been successful in retrieving the queen, then this will be fairly quick providing that the disturbance was minimal. If the colony was large, and difficult to remove there is potential for the queen to be killed by the colony themselves, almost blaming her for the current state of affairs. I have seen this happen on a small number of occasions. If you were unable to find your queen, or she was killed during the removal, your colony will take 24-48 hours to settle, in which time they will start the process of creating a new queen. Of course, this is providing you have collected the correct ingredients for this to take place, as outlined previously in the section above.

A busy quarantine apiary at the end of the day. At some points of the bee removal season, and despite much planning, it can get a bit hectic and tight for space, as can be seen here. More stands were brought to site after this picture was taken. Photo by Clive A Stewart

It is extremely important that you employ strict hygiene protocols throughout the entire process and identify one of your apiaries as a quarantine apiary helps to ensure that any outbreaks of notifiable diseases are recognised as early as possible.

Part 5

Planning a Colony Removal

Part 5 – Planning a Colony Removal

Case Study 1 – Cut-Out

I picked up the phone to hear the voice of a rather stern, but well-spoken gentleman. "Are you the bee-man?" and before I could answer "More to the point can you take bees from out of buildings!". I informed the gentleman that he had indeed contacted the person he had been looking for. He went on to describe a recent swarm that had gone into an air vent on the front elevation of his property and that a number of bees were looking at other vents in the vicinity. This sounded like a recent swarm, and one that could be dealt with relative ease.

A date and time was agreed for an initial survey. From my recollection of the morning of the survey, it was pleasantly mild, and the bees were very active. It was clear that the bees had, as the gentleman assumed, utilised one of the vents at the front of the property, and that there appeared to be a number of bees looking slightly disorientated.

Thermal imaging scans were completed both inside and outside in order to try and pinpoint the colony's position with inside the building. Was it a long established colony? Had it gone into the wall cavity? Was it in the roof space? Under the loft floor? In short, it had the potential to be just about anywhere. The thermal imaging was inconclusive and further survey techniques were employed. An endoscope camera was inserted into the air vent but come up against some brick work. However, there was a gap up the side of the brick work where the eaves were sat and the bees were going past this looking as though they were going into the loft space.

I requested access to the loft area which was heavily insulated, coupled with some very intricate roof structure. I was able to get close to the location and locate part of the comb inside the loft space though only a small amount of fresh comb was visible, so it was not possible to determine just how large the colony was going to be. What was clear was that we were going to require scaffolding and tiles were going to need removing to enable us to expose the colony.

A report was written up along with recommendations, along with a quote for the works. Some days later the quote was accepted and a date was scheduled. The scaffold team came in on day one, and had some difficulty in erecting the scaffold around the garden as the client was a keen gardener, and clearly very proud, and protective of their plot. The team explained how they would be careful around the flower beds, and work commenced in a cautious and attentive manner. All the necessary safety features, such as the toe boards & safety gates were attained, and the scaffold was signed off.

The home of the bees – photograph by Clive A Stewart

Bees coming and going from the air brick – photograph by Clive A Stewart.

The scaffold in-place and work is ready to start – photograph by Clive A Stewart.

Day two was the day of removal and a warm sun made it all the more pleasant. Bees were actively bringing in nectar and pollen, and the activity around the other air vents had ceased. All the tools and equipment were taken up to the work area, including the transport box. We began by removing the first two rows of tiles. There was an abundance of bird nesting material. It was noted at the time of the survey that there was no bird activity, and this was confirmed by the empty nests. I should mention that if these had of been identified as being occupied at the time of the survey the recommendation would have been to remove the bees after any chicks would have fledged, and equally the same if there had been nesting birds moved in during the time line since the survey and we had found them during the opening up we would have simply stopped and replaced the tiles leaving them till a time when the chicks would have left the nest. A decayed wasp nest was also present in this area, though there wasn't much left of it.

Bird nesting evidence – photograph Clive A. Stewart

Roof tiles off exposing the honey bee colony and an old wasp nest – photograph by Clive A. Stewart

Removal box ready to take the bees – photograph by Clive A. Stewart

On removing the tiles, the laths required cutting as did the felt in order to peel it back in a neat and tidy fashion. It was clear from this that we had not quite removed enough and a third row was removed in order to gain a little more room to gain easier access. The felt was slowly peeled back to expose the colony to the open air. The bees came into view along with a second much more intact wasp nest. Thankfully this too was uninhabited. Lord only knows what commotion, and complexities that might have presented had it been occupied.

It was clear that had this honey bee colony been left for any length of time they would have easily filled the void space quite rapidly as they were already starting to build their comb around the roof trusses. They were distinctively orange in colour, and extremely calm showing very little in the way of any defence traits.

Two pieces of the fresh comb closest to me dropped from the felt, this did not arouse any retribution. These were placed with bees into the transport box, and the slow and methodical removal of the individual combs were removed, and framed using the elastic band to frames method. Only because I had simply run out of comb clamps, which are my preferred method.

The felt peeled back to reveal the bees and the old wasp nest – photograph by Clive A. Stewart

The bees remained calm throughout the entire ordeal and did not appear stressed at all. Of course we may have been looking at a completely different story if the colony had been established for any length of time, and the number of bees had been greater. The situation may not have been so pleasant and the use of a bee vac may have been needed, had the number of bees been more typical of such a colony. Whilst the colony was not the largest it was noticeably productive, and headed by the most beautifully marked queen.

Oh yes you did indeed read that correctly, and it is not uncommon particularly in the early part of the year to come across marked queens. Now these queens don't paint themselves in pretty colours, as there are no mirrors in the nest, and workers cannot hold paint brushes. No these are in fact the unfortunate result of a beekeeper who has missed the swarm preparation signs within the colony he is managing. There can of course be a number of reasons a managed colony has been allowed to swarm, the beekeeper could be ill and is simply not able to manage his colonies as they should, or there may be other such mitigating factors. However, there is a worryingly noticeable increase in hobby beekeepers that watch a you tube video or two and delve head first without suitable training. Excited with the glamour and romance of idealistic notions of becoming billionaire honey barons they miss the fundamental basics of responsible beekeeping. When I hear of such notions from potential new comers to the world of beekeeping and bee removal for that matter, I remind them that it is paramount above all else to seek suitable training, and understand that there is an equal level of responsibility to keeping bees as there is to keeping any other livestock. To under estimate this is costly to both the new beekeeper, and any nearby neighbours.

With the colony all framed up, and placed into a suitable transport box, I placed the box in the flight path of the air vent. I then went about proofing the inside of the roof space while the bees re-orientated to their new home. This was done by filling the void space in that area with loft insulation in order to prevent the bees re-establishing themselves. I then temporarily replaced the roof tiles back in place to give the roof some level of weather protection until the builder returned 24 hours later. This is to allow any robbing bees and nuisance wasps to clear up any spilt honey residues and reduce the risk of the builder being stung by those carrying out the clean-up operation. I like to keep my sites as clean and tidy as I can, but there is inevitably always a drip of honey somewhere and the robbers will find it. Thankfully there was not a lot on this removal, and they had cleared most of it up throughout the day.

Can you spot the marked queen? Photograph by Clive A. Stewart

Once the bees had settled and all the flying had stopped at dusk, we shut up the entrance and transported the colony over to our quarantine apiary more than three miles away. This is extremely important in ensuring that no bees return to the site the following day. Here they remain to be monitored for notifiable diseases, and are checked by the local seasonal bee inspector usually within six weeks of being on site. As long as the colony is fit, it is then moved in to a production apiary to be managed for honey gathering.

I returned the following day with the builder to reinstate the roof properly, repairing the felt, securing the roof lathes, and bedding the tiles on the gable end. The scaffold was removed that afternoon and the occupants had their home all to themselves again. Their relief was quite obvious when the work was completed, and they were so pleased that the colony had been saved in the process.

The remaining stragglers being called to their new hone at the end of the day –
Photograph by Clive A. Stewart

Case Study 2 – Trap-Out

It was mid-afternoon when I got one of those hysterical calls, and the person on the other end was a little difficult to understand. Customer service skills are invaluable at this point and calming the situation down so that a sensible conversation can be held without the frenzied explanations of the great apocalypse that has besieged their property is not going to have their world collapsing around them, is a true measurement of patients.

I should mention at this point that the reader may be surprised to discover that not everyone likes bees, and that some people have true phobias about any insect that has stripes and has the potential to sting them. These individuals share no support for the dedication & love you show to your chosen profession and can be exceptionally testing at times. Thankfully on this occasion the lady concerned was quite understanding of her situation. However, this was only once it was explained that the situation was not uncommon and that the issue could be resolved. Sadly, it was not quite as straightforward as I had hoped.

The lady in question lived with her family not a million miles away from my front door, and it was agreed that a survey could take place that very afternoon. On arrival the usual meet and greet was conducted, and observations of the activity noted. The bees were bringing in copious amounts of pollen and nectar. This is a sure sign that the colony is well established and has been there for a sufficient amount of time in order to have built comb to store the abundant supplies being brought in, and quite possibly raising brood too. Whilst the family had only been away for seven days the colony was clearly beyond the realms of a forced abscond, and it was obvious that more drastic action would be needed.

It was a red sand stone cottage of considerable age with extremely thick walls, a new extension had been built on to the old part of the building, and where the two buildings met, the new was slightly forward of the old which created an inverted corner, and as the block work was of informal stature, they formed a gap where it did not meet perfectly. The bees had taken full advantage of this, and it appeared that they may potentially be in the floor or ceiling of either of the two properties.

The bee activity was in the corner just above the meter box – photograph by Clive A. Stewart

An internal survey of the areas concerned was undertaken. Floors, walls, and ceilings were checked, and scanned with the thermal imaging camera, but alas no heat signature could be detected.

Because of the type of structure and its age, there was reluctance from myself and the family to dismantle any part of the building relying solely on guess work alone. The old stone walls were as thick as 18 inches in places, possibly more, and were unlikely to have a cavity as such. It was decided that the only potential place they could possibly be, was between the two houses, but because no heat signature could be found it was decided that a trap out would be the most favourable on this occasion.

A report was written up along with the aforementioned recommendations, a plan of works, and a quotation. There was no need for scaffolding or high-level access equipment, as the entry point was fairly low and it could be reached of a small set of steps, and unlike cut out removals trap outs require less equipment. Each segment of the work would only be for short periods of time too.

Stage one began with the creating the surrogate colony that was to be sited at the property. Giving the established colony of disorientated bees-to-be, somewhere to re-orientate to. This was made up using a wooden nucleus hive with two frames of brood in all stages and a frame of food to help sustain the surrogate colony whilst it got itself established.

It was purposely created 'queenless', in order to entice any potential emerging queen from the target colony to take up residence in the supplied surrogate colony. If this was conducted with a queen inside the surrogate colony any emerging queen from the target colony would potentially disregard the new home as already being governed and would likely be rebuffed by the tenants when attempting to gain access.

Once the surrogate colony had been created, they would require siting. The surrogate colony needs to be placed as close to the entrance of the target colony as is feasibly possible but allowing enough room for the trapping cone to be secured over the entrance.

The siting of the surrogate colony is vitally important and is to give the disorientated bees to be a place to re-orientate to. In turn, this provides those bees with a place to unload their incoming supplies, which they then offer to the guard bees on their return as a bribe to get in, and after deciding that there is nowhere else to go.

Sighting of the cone and the receiver colony – photograph by Clive A. Stewart

On this occasion securing the nucleus was done by simply fitting two shelf brackets in place large enough to take the weight of the wooden nuc box. "Why did you not use a polystyrene nuc?" I hear you cry. The polystyrene nucs that we use and most of those available only have the little hole at the front or either end. These would potentially slow down the acceptance of bees getting into the surrogate colony. Put simply the larger entrance will allow more bees in at a time and it is harder for the surrogate colony to defend.

111

The brackets and cone in position just above the meter box – photograph by Clive A. Stewart

Securing the trap out cone is stage two and was not an easy task on this particular site, due to the inverted corner. I always choose to use a face plate to mount the cone to. This is normally a piece of 9-12mm scrap board usually 200mm x 200mm with a 65-70mm hole in the middle. Fitted into the hole is a wire gauze cone that is formed to create a 9mm hole at its narrowest point and bound in frame wire for stability and to prevent the cone opening up. It is then stapled in place on to the board. This can then simply be either screwed or stuck into place onsite.

On this occasion we had to put a 90-degree angle on the face plate to enable us to fit it over the entrance that was in the crevice of the inverted corner, and because the surface was so uneven it was decided that it would be screwed into place. Once the cone was secured, we had to sponge the gaps around the face plate in order to stop the foraging bees of the target colony gaining entry back into the building.

A close up of the inside of the cone – photograph by Clive A. Stewart

Regular return visits over the following three to four days is vitally important to the success of a trap-out. This because the determined foragers want to get back in and will literally look everywhere in order to find a way back into the colony.

I have witnessed them find entrances over two metres away in order to regain entrance to their nest site. So always have plenty of blocking up material to hand. Over the coming days the bees from the target colony are bled from the sand stone wall of the cottage, and they slowly merge with the surrogate colony.

As the days go by, the return visits can be less frequent, leaving a day or two in between. I usually monitor the number of bees leaving the cone as a guide to how well the trap out is going. In the early days there will be a lot of activity around the base of the cone, where bees are trying to work out a way back in, and as the colony wains it is quite apparent that there are less bees coming from the cone and more bees coming from the surrogate colony.

It is quite a satisfying feeling when you return to the colony, and you see no bees around the base of the cone and all the activity coming and going from the surrogate colony entrance it is at this point you can consider removing your trapped out bees and returning the site back to the client.

During the surrogate colony removal process try not to disturb the colony to much. However, remember that after the first seven days of being in situ you will need to do an inspection and remove any queen cells from the surrogate colony. You can potentially do this before they go to site, but I prefer to do it on site, purely only for the reason that the colony will have bulked up over those first initial days, with the returning foraging bees of the target colony. In contrast, if the inspection is done at the home apiary beforehand, those numbers will only dwindle.

Back to our trap out at the cottage, with the cone in place the bees began to do what bees do when they are prevented from getting in where they normally would, and began searching other areas. Over the following two or three days we blocked up a number of points including brickwork meeting the eves, and the electric box to the side. By day five, they were starting to use the box more regularly, though there were still a large number of disorientated bees. This pattern continued for well over a week.

By the end of day fourteen the bees that were exiting the cone were becoming noticeably less and site visits were down to every three days. At the end of week three there were no bees leaving the cone with just a very small number of ten or twenty bees still trying to orientate to the original entrance. On the last day I waited till dusk when all the flying had ceased, sealed up the entrance and removed the bees to relocate them in the quarantine apiary.

The need to quarantine these bees is equally as important as a cut out removal if not somewhat more as you are unable to examine for any brood diseases. I did not see the queen leave and I have yet to actually witness this happen. They don't always take to the new colony. However, on this occasion it was a successful removal, and the queen was found to be laying on the following inspection.

As for preventing the next passing swarm taking up residency, it was simply a case of sealing back up the entrance and those areas we found them to be re-entering during the process, such as the electric box, and the other brickwork.

A bee leaving the cone – photograph by Clive A. Stewart

Case Study 3 – Swarm Collection

As a bee removal specialist, I rarely get calls for swarms. However, due to the nature of this particular incident we got an interesting call for bees in a building which turned out to be a swarm.

Allow me to explain, the local swarm catchers know me well and have seen the work that we do, and as a consequence of this whenever they receive a call for bees in a building, they forward it straight to me.

On this occasion the client had described that a swarm of bees were getting into their building. Not describing what sort of building, the swarm coordinator at the time immediately advised that they need to speak to me, passed on my number, and hung up. I would usually oblige with the same, and return the favour if the caller has a swarm hanging from a tree or shrub in the garden. I do this as hobby beekeepers will go and collect swarms from these situations for free or a small donation towards fuel, whereas I'm a business and would charge a minimum call out fee which would also include time spent on site and travel.

The BBKA (British Beekeepers Association) actually have a list of hobby beekeepers that do this. Now it must be noted that beekeepers that are affiliated to the BBKA are insured only to collect swarms and not remove bees from buildings commercially. I stress this, because the swarm coordinator in all fairness did the correct thing as the bees had actually got inside the building, but as you will see it was plausible for the swarm collection to have been done by them.

The call came in, and the client described that they had bees getting into their building. Being slightly alarmed they refrained from describing what type of building it was but did describe the bees to be low down and would not require access equipment. Puzzled by what was going on, I arranged for a survey that afternoon as I was at a loose end and would enjoy a run out fitting in some beekeeping activities on the way home.

I made my way across the county and arrived at some rather large and fanciful iron gates to be let in by security. Driving up the driveway that seemed to go on for ever, passing a deer park and some rare breed cattle I eventually get to the front doors of a big house. I was just about knock on the door when an elderly gentleman, dressed in country attire, came around the corner. "Are you the bee man?" I acknowledged that I was the person he was looking for. "This way mate… They're round ere". I instantly recognised his voice as the person who had called me earlier, and I did as was instructed. Taken through the well maintained gardens I was lead into a walled garden. I was expecting to see bees going in and out of a hole in the big mansion, but low and behold it was a swarm that had landed on a recently renovated Victorian greenhouse.

The swarm cluster on the greenhouse – photograph by Clive A. Stewart

Well, I suppose they were technically correct, the bees were getting in to the building as there were some inside as well as the main cluster being on the outside. It turned out that the gentleman that was showing me to the spot was indeed the gardener. The greenhouse was where he would have his lunch and tea breaks, and he described nearly walking into the bees when he was coming in for his afternoon tea. He hadn't seen it arrive as he had been working on the other side of the house far from where the greenhouse was situated.

Bees inside the greenhouse – photograph by Clive A. Stewart

It was a mere stroke of luck that I had decided to take some equipment with me for beekeeping activities on the way home. However, it was only a limited amount of equipment, and the swarm was of a good size. There was no need for the survey kit at all as you could see exactly what was going on. So I went off to the van to collect the necessary equipment. Sadly, all I had was an old nucleus box that was going to be used to split a colony on the way home. The nucleus was potentially a bit on the small side, so I decided to remove all the frames except one of drawn comb and attempt to get the swarm in it.

The next problem would be how best to get them in. Located on the upper corner of the building at the bottom of the roof where the gutter would usually be, they could not be shaken from their position in the same way you might collect one from a branch. It could have been brushed into the box, but not an easy task. The obvious way would be to allow the bees to use their natural instincts and allow them to walk up into the nucleus, though placing a wooden box on a slope was going to be a little difficult. The gap between the joists of the greenhouse roof were large enough to site the nucleus and a prop was used to keep it in place.

The smoker prepared with the usual dead grass and shavings which when lit produced a rather heavy but cool smoke. The bees responded slowly to the smoke which encouraged them to move upwards towards the entrance.

After a few minutes of this the bees began to enter the nucleus. A little unsure one or two of them went in and came back out, but as more and more of them slowly made their way up more and more began to consider it a suitable home, and it wasn't long before a number of the workers were fanning with their tails in the air, and the remaining bees started to march in.

The colony soon engulfed the nucleus, so much so that it was advantageous to prop open the cover board in order to allow the crowd in to their temporary home. It took them around half an hour for the bulk of them to get in.

The empty nuc placed above the swarm cluster – photograph by Clive A. Stewart

The nucleus was left till dusk when all the bees had ceased flying and the colony was settled. On my return I closed the entrance, with a sponge block, and loaded them into the van to drop at the nearest apiary which was a little more than 3 miles away from the Victorian walled garden.

I didn't see the queen during the collection, perhaps because of the long discussions with the gardener on the subject of bees and his vegetable patch, but when I went to upgrade the bees' living accommodation to a full hive, I saw that she was marked quite clearly with a yellow spot.

Another beekeeper had lost a fantastic colony as they went on to produce a tidy amount of honey for that year.

The bees have all entered the nuc – photograph by Clive A. Stewart

Part 6
Summary & Conclusions

Part 6. Summary & Conclusions

Summary

This book gives the reader an insight to the basic elements of what is rapidly becoming an industry all of its own. It outlines key knowledge points for further research, and should not be considered a substitute for formal training. I like to read other beekeepers' ideas on this diverse subject, though I have very little time these days so instead I choose to watch you tube. I watch both the good and the bad, and I like to listen to the podcasts in equal measure, usually while making frames in the work shop during the cold winter months. However, in exactly the same light as this book these should not be considered as an alternative for formal training but used as one form of research.

I have had an admiration for bees for a long time, long before I took up beekeeping, and long before it even became trendy. Beekeeping has certainly become very trendy. In the last decade I have seen a vast number of enthusiastic people come and go in the romance of the smog filled aura floating round in the back garden tending their friendly colony in its little white house and collecting a jar or two of their very own honey as a prize for their good deed. However, when the reality kicks in it is often a much more a love hate relationship. The new owners often still love their honey but recognise that it isn't quite what they imagined, and grumpy colonies are left to fend for themselves. Couple these individuals with the people I group as the natural beekeepers who feel keeping bees in logs in the back garden, allowing them to get on with it is the only way bees should be kept, and we have the perfect recipe for an apoplectic plague of swarms.

Forgive me I'm allowing my passion to exaggerate, but there has been a noticeable growth in troublesome swarms steadily increasing with the popularity of beekeeping, and it should be said that the vocation of keeping bees is potentially no longer about saving the bees, but more about responsible beekeeping. I hope it has become clear to the reader by now, if they have managed to get this far, that the cost implications involved in removing bees from a building is not going to be a small sum, or a simple case of arriving in a bee suit and puffing a little bit of smoke in here and there or smashing a couple of bricks out in the hope that it is just a little cluster of bees.

On more than one occasion I have come up against attempts by hobby beekeepers that have got part way into situations and have realised that they have seriously made a big mistake. I do try to encourage the client to bring back the beekeeper. They have always reluctantly declined. Going to tidy up someone else's mess is never a pleasant experience. I recall one particular incident where a roofer was being disrupted by bees in a chimney. He found his client a local beekeeper who proclaimed to be qualified in the art of beekeeping and could remove these bees with the help of a builder.

The roofer offered his services, and they began the works the following day. The beekeeper who was a young man brought along his father for assistance and a spare bee veil for the roofer who was going to open up the brickwork. It transpired that on the removal of the first few bricks the roofer was slung a number of times, the young beekeeper also took a sting or two but undeterred he continued. He managed to remove four fins of comb placed them on the top of the chimney and must have decided it was too big for him to complete. He informed his client that he had sufficient bees in his transport box and retreated from site never to be seen again. The bees were doomed from that point on as they were attacked by marauding bees, and by the time I arrived they were found to be extremely exhausted from attempting to defend their stores. Bee removal is not for the faint hearted, and it is certainly not away of obtaining free bees. Beekeeping knowledge alone is simply not enough to be able to carry out this sort of work, and all aspects need to be considered when looking at those situations that require taking apart someone's home or business in order to relieve them of their bee problem.

Questions & Answers

Why has honey bee removal become so popular? SR to CS

The reason bee removal has become the go to method for dealing with honey bee situations in particular is mainly down to the vague and incoherent labels on the chemicals available for the use against bees. They are not clear and are open to misinterpretation, often leaving the pest controller open to persecution, and in worst scenario case prosecution.

This inadvertently poses a risk to the human food chain as foraging bees can come into contact with those insecticides deemed suitable. The construction of a honey bee nests does not allow for successful application of modern day chemicals as more often than not the chemical simply does not reach the core of the nest. These have become the main reasons why pest control companies refuse to treat colonies today, and why live bee removal has become so popular.

Can anyone become a bee remover? SR to CS

Well I suppose in the same way as anyone can become an astronaut? Yes, you can, but only with the correct mind set, equipment, and suitable training to prove competency. It is all very well having a little knowledge of bees, putting on a suit, turning up with your chosen bee vac, and believing you can go knock a hole in someone's castle, but get it wrong and it could cost you a lot more than a trip to court.

Cutting corners on safety, can potentially see you putting your own and other lives at risk. Not investing in the right equipment can see you struggle with the simplest of tasks. Astronauts don't just read a book on how to fly a spaceship and turn up at Nasa and ask to have a go do they?

Multiple colonies in a single location is becoming more and more common. According to Professor Tom Seeley this is not supposed to be the natural instinct of honey bees in the wild. Do you have any thoughts on why this might be happening? CS to SR

It is my understanding that when western honey bee colonies swarm that the tendency is for them to initially cluster withing 20 metres or so and then to move to a new nest site over a kilometre away. The practice of having lots of colonies very close together is an unnatural state for the bees to be in. I personally do not have an answer to why this is becoming more frequent. However, in a given location, it might be that the best nesting cavity, or even the only nesting cavity available, may be one in close proximity to another colony. The bees may not have had a choice. In short, I don't know the answer. Perhaps we should ask Professor Seeley?

Conclusions

1. Bees in buildings are not free bees

2. Do not attempt a bee removal before getting the appropriate training

3. Do not work on bee removal from a ladder

4. Do not use the honey from a bee removal for human consumption

Further Reading

▶ *Honeybee Democracy*, published by the Princeton University Press. Written by Professor Thomas Seeley. This book tells the story of swarming and how the bees choose a new nest site. I would recommend this book to anyone with an interest in honey bees (SR)

▶ *The Lives of Bees: The Untold Story of the Honey Bee in the Wild.* Written by Professor Thomas Seeley. This book tells the story of what wild colonies of bees are doing. There is a very interesting section called 'Darwinian Beekeeping' at the end of the book. (SR)

▶ *Practical Beekeeping.* Published by The Crowood Press Ltd, 1997. Written by Clive de Bruyn. A comprehensive book on keeping honey bees. (CS)

▶ *Silent Earth: Averting the Insect Apocalypse.* Published by Vintage Publishing. Written by Dave Goulson. An interesting and terrifying view of the future. (SR)

"The bee's life is like a magic well:
the more you draw from it, the more it fills with water"

- Karl Von Frisch, Bees: Their Vision, Chemical Senses and Language

Karl von Frisch was awarded the Nobel prize for Physiology or Medicine in 1973 for his research work on honey bees.

What we take from these words is that the more we learn about bees, the more, we realise, there is to learn.

About the Authors

Clive Stewart has been keeping bees for 23 years and began removing bees back in 2008 at the time working as a gardener. Two years later saw him retrain and move into professional pest control. At which time there was great controversy in regard to honey bees and how they were treated.

Since then, he has been pivotal in the development of bee removal work within this industry and has been influential in many of the changes bringing it to its recognised status today. He is also responsible for the creation of the UK Bee Removers (UKBR) in 2020. The first organisation to be recognised specifically for the purpose of the small but growing industry of bee removal. Originally set up to assist the public in finding a bee remover in their local area it has since also worked towards highlighting the importance of safe working practices.

Stuart A. Roberts has been keeping bees for 13 years, at the time of writing. He is a British Beekeepers Association (BBKA) Master Beekeeper, becoming qualified in 2018.

He keeps locally adapted mongrel honey bees in Staffordshire across eight out-apiaries, running about 40 colonies. He is an active member of the BBKA Examinations Board and a trustee for the International Bee Research Association (IBRA).

Ingram Content Group UK Ltd.
Milton Keynes UK
UKHW050052120523
421599UK00005B/46